MRI: Connecting the Dots

A start to concepts

Online at: https://doi.org/10.1088/978-0-7503-1284-4

IPEM–IOP Series in Physics and Engineering in Medicine and Biology

Editorial Advisory Board Members

Frank Verhaegen
Maastro Clinic, The Netherlands

Carmel Caruana
University of Malta, Malta

Penelope Allisy-Roberts
formerly of BIPM, Sèvres, France

Rory Cooper
University of Pittsburgh, PA, USA

Alicia El Haj
University of Birmingham, UK

Kwan Hoong Ng
University of Malaya, Malaysia

John Hossack
University of Virginia, USA

Tingting Zhu
University of Oxford, UK

Dennis Schaart
TU Delft, The Netherlands

Indra J Das
Northwestern University Feinberg School of Medicine, USA

About the Series

The Series in Physics and Engineering in Medicine and Biology will allow IPEM to enhance its mission to 'advance physics and engineering applied to medicine and biology for the public good'.

This series focuses on key areas including, but not limited to:
- clinical engineering
- diagnostic radiology
- informatics and computing
- magnetic resonance imaging
- nuclear medicine
- physiological measurement
- radiation protection
- radiotherapy
- rehabilitation engineering
- ultrasound and non-ionising radiation.

A number of IPEM–IOP titles are being published as part of the EUTEMPE Network Series for Medical Physics Experts.

A full list of titles published in this series can be found here: https://iopscience.iop.org/bookListInfo/physics-engineering-medicine-biology-series.

MRI: Connecting the Dots

A start to concepts

Dee Wu
*Department of Radiological Sciences, University of Oklahoma
Health Sciences Center, Oklahoma, USA*

IOP Publishing, Bristol, UK

© IOP Publishing Ltd 2023

All rights reserved. No part of this publication may be reproduced, stored in a retrieval system or transmitted in any form or by any means, electronic, mechanical, photocopying, recording or otherwise, without the prior permission of the publisher, or as expressly permitted by law or under terms agreed with the appropriate rights organization. Multiple copying is permitted in accordance with the terms of licenses issued by the Copyright Licensing Agency, the Copyright Clearance Centre and other reproduction rights organizations.

Certain images in this publication have been obtained by the author from the Pixabay website, where they were made available under the Pixabay License. To the extent that the law allows, IOP Publishing disclaims any liability that any person may suffer as a result of accessing, using or forwarding the images. Any reuse rights should be checked and permission should be sought if necessary, from Pixabay and/or the copyright owner (as appropriate) before using or forwarding the images.

Certain images in this publication have been obtained by the authors from the Wikimedia website, where they were made available under a Creative Commons license or stated to be in the public domain. Please see individual figure captions in this publication for details. To the extent that the law allows, IOP Publishing disclaim any liability that any person may suffer as a result of accessing, using or forwarding the images. Any reuse rights should be checked and permission should be sought if necessary, from Wikipedia/Wikimedia and/or the copyright owner (as appropriate) before using or forwarding the images.

Permission to make use of IOP Publishing content other than as set out above may be sought at permissions@ioppublishing.org.

Dee Wu has asserted his right to be identified as the author of this work in accordance with sections 77 and 78 of the Copyright, Designs and Patents Act 1988.

ISBN   978-0-7503-1284-4 (ebook)
ISBN   978-0-7503-1285-1 (print)
ISBN   978-0-7503-1858-7 (myPrint)
ISBN   978-0-7503-1286-8 (mobi)

DOI   10.1088/978-0-7503-1284-4

Multimedia content is available for this book from https://doi.org/10.1088/978-0-7503-1284-4.

Version: 20230201

IOP ebooks

British Library Cataloguing-in-Publication Data: A catalogue record for this book is available from the British Library.

Published by IOP Publishing, wholly owned by The Institute of Physics, London

IOP Publishing, No.2 The Distillery, Glassfields, Avon Street, Bristol, BS2 0GR, UK

US Office: IOP Publishing, Inc., 190 North Independence Mall West, Suite 601, Philadelphia, PA 19106, USA

For residents, technologists, scientists, and/or anyone curious.

Contents

Preface	xiii
Acknowledgement	xv
Author biography	xvi
How to use this book	xvii
Introduction for residents, technologists, scientists, and for anyone curious	xxiii

1 Five big ideas for MRI — 1-1

1   Introduction — 1-1
1.1   Waves — 1-7
    1.1.1   Two types of waves — 1-8
    1.1.2   Electromagnetic (EM) waves — 1-9
    1.1.3   Mathematical properties of transverse waves — 1-9
    1.1.4   Understanding energy of waves in relation to frequency of waves — 1-10
    1.1.5   Decay/damping of waves — 1-12
    1.1.6   'Phase versus frequency' — 1-14
    1.1.7   Understanding frequency spectrum and frequency bandwidth — 1-16
    1.1.8   Adding different waves together (a form of synthesizing waves) — 1-20
    1.1.9   Fourier transform — 1-23
    1.1.10   Summary — 1-27
1.2   MRI as a water image — 1-28
    1.2.1   MRI utilizes water — 1-30
    1.2.2   Electric dipoles — 1-31
    1.2.3   Water shifts — 1-32
1.3   Multiple looks (contemplating image weighting and the multiple appearances possible in MRI) — 1-33
    1.3.1   How do we achieve multiple looks/multiple weightings — 1-34
    1.3.2   Multiple looks and weightings as the analogy of different dimensions — 1-35
1.4   Interlude and segue into concepts of electricity and magnetism — 1-37
    1.4.1   Checkpoint up to this part of the chapter — 1-37
    1.4.2   Review of basics of electricity and a little bit of history — 1-37
    1.4.3   Electricity and magnetism and the world around us — 1-40
    1.4.4   Learning objectives — 1-41
    1.4.5   Michael Faraday and the EM field — 1-41

1.5	Changing fields alter E&M	1-46
	1.5.1 Concept number 4, changing electric field → makes magnetic field, and changing magnetic field → makes electric field, which we can also provide names to as Faraday's law ($\Delta M \to E$) and Ampère's law ($\Delta E \to M$)	1-46
	1.5.2 Putting together $\Delta M \to E$ and $\Delta E \to M$—suggested experiment to run in class	1-47
	1.5.3 Electromagnets	1-48
1.6	Dipoles and precession	1-50
	1.6.1 What is a dipole?	1-50
	1.6.2 Magnetic dipole	1-51
	1.6.3 Precession	1-54
	1.6.4 Putting together the main idea: protons have 'spin' and thus angular momentum, so they have ability for precession in the electrical sense, just like gyroscopes have the ability to precess in the mechanical domain	1-56
	1.6.5 Knowledge check for the magnetic dipole	1-58
I-A	Appendix of chapter 1	1-59
	References	1-72

2 Hardware: five important components and beyond — 2-1

2.1	Introduction	2-1
	2.1.1 Ten reasons why it may be good to start with a perspective on MRI hardware	2-3
	2.1.2 Yield bubble	2-4
	2.1.3 Brief overview of the anatomy of the scanner	2-4
2.2	Main magnet	2-5
	2.2.1 Different MRI magnet types	2-6
	2.2.2 Suggestions on DLS and how to think about the main magnet field	2-7
	2.2.3 Location of the main field	2-8
	2.2.4 Role of the solenoid	2-9
	2.2.5 Summary to this point	2-10
	2.2.6 Alignment of spins	2-11
	2.2.7 Fun-fact bubble: Analogy for the two states in MRI for 'parallel' and 'antiparallel' spins	2-12
	2.2.8 Scaffolding bubble	2-14
	2.2.9 Permit precession	2-15
	2.2.10 Precession revisited and Larmor frequency	2-15
	2.2.11 Summary of the main field	2-16

2.3	Shim coils	2-17
	2.3.1 The goals of shimming (reduce unintended gradients)	2-17
	2.3.2 Scaffolding bubble	2-18
2.4	MRI gradient coils	2-19
	2.4.1 Summary	2-24
2.5	Radiofrequency (RF) transmitter	2-25
	2.5.1 Roles of the RF transmitter	2-25
2.6	RF receiver	2-28
2.7	Scaffolding bubble	2-30
2.8	The briefest of introductions to k-space (traversing frequency space)	2-32
2.9	Summary and looking ahead	2-37
	Appendix 2.A	2-37
	References	2-41

3   Basic building blocks for MRI   3-1

3.1	Introduction	3-1
3.2	Contrast and spatial resolution	3-4
	3.2.1 Contrast	3-5
	3.2.2 Spatial resolution	3-7
3.3	Relaxation (what, how, and when?)	3-10
	3.3.1 What does weighting look like?	3-11
	3.3.2 Scaffolding bubble: Synthetic imaging, creating images from weighting parameters, and synthesis	3-11
	3.3.3 How do TR and TE affect weighting of T1 and T2?	3-12
	3.3.4 How does relaxation work? Examining longitudinal and transverse magnetization	3-15
	3.3.5 T1 as regrowth (dissecting the details)	3-15
	3.3.6 Scaffolding bubble, visualizing T1	3-17
	3.3.7 T2 as decay	3-18
	3.3.8 When you understand weighting and how it works, where do we go next with relaxation, clinically?	3-21
	3.3.9 Application of imaging weighting (given TE, TR parameters)	3-28
3.4	Including inhomogeneity T2* (into T2)!	3-30
	3.4.1 What does T2* effect look like on images	3-30
	3.4.2 What is meant by susceptibility?	3-31
	3.4.3 T2 versus T2*	3-32
	3.4.4 How does inhomogeneity term contribute?	3-34
	3.4.5 Summary	3-34

3.5	Table of T1/T2 and physical values	3-35
3.6	Summary	3-36
	References	3-36

4 The inside details of MRI — 4-1

4.1	Introduction	4-1
4.2	Free induction decay (FID)—your first signal	4-2
	4.2.1 A simplistic recipe for producing an FID	4-4
	4.2.2 Summary: two key memorization points for FID	4-5
	4.2.3 Scaffolding Bubble: Motivations for waves revisited	4-6
4.3	Pulse sequence diagrams	4-8
	4.3.1 Let's look at the 'tusk/spear' of the elephant, i.e., the slice encode	4-13
	4.3.2 Looking at the 'ear/fan' of the elephant, i.e., the frequency encode	4-17
4.4	Scaffolding box: Steps of what is happening in frequency encoding	4-18
	4.4.1 Summary: two key memorization points for frequency/read encoding gradient	4-21
	4.4.2 Looking at the 'legs/pillars' of the elephant, i.e., the phase encode	4-22
	4.4.3 Summary: two key memorization points for phase encoding gradient	4-23
	4.4.4 INSTRUCTOR TIP	4-24
	4.4.5 Summary of parts of the PSD all together	4-25
	4.4.6 Scaffolding bubble on TE and TR	4-27
4.5	Spin echo	4-27
	4.5.1 Applied bubble: pituitary gland	4-29
	4.5.2 History bubble: runners on a track analogy for the spin echo	4-32
4.6	Summary	4-34
	Appendix 4.A	4-35
	References	4-42

5 Getting serious with MRI — 5-1

5.1	Excitation with RF transmit	5-2
5.2	Traversing with k-space and qualities of the image	5-8
	5.2.1 Retrieving spectrum in multiple dimensions and the center versus periphery k-space	5-9
	5.2.2 Traversing k-space with gradient structure of the pulse sequence diagram	5-13
	5.2.3 Scaffolding bubble	5-14

5.3	On the receiver side	5-15
	5.3.1 Signal-to-noise ratio (SNR)	5-16
	5.3.2 Contrast-to-noise	5-17
	5.3.3 Bandwidth (receiver): a listening perspective	5-17
5.4	Summary	5-20
	References	5-22

6 Three tradeoffs in MRI (Clinically relevant)   6-1

6.1	Introduction	6-1
6.2	Factors that change time	6-1
	6.2.1 Repetition time (TR)	6-3
	6.2.2 Number of phase-encoding steps and number of averages effects change on time	6-4
	6.2.3 Knowledge check	6-6
6.3	SNR tradeoffs	6-7
	6.3.1 Spatial size and abundance	6-10
	6.3.2 Averaging	6-13
	6.3.3 Scaffolding bubble: signal averaging	6-14
	6.3.4 SNR changes based on time-related effects (sampling time/receive bandwidth) and relaxation effects	6-16
	6.3.5 Knowledge check	6-20
6.4	Tradeoffs with SAR (energy)	6-21
	6.4.1 Fun-fact bubble	6-22
6.5	Knowledge check	6-25
6.6	Summary of MRI tradeoffs	6-26
	References	6-27

7 MRI Artifacts (clinically relevant)   7-1

7.1	Introduction	7-1
7.2	Starting with initial concepts on artifacts	7-2
	7.2.1 Distortion (geometric)/susceptibility artifact	7-2
	7.2.2 Partial voluming	7-4
	7.2.3 Multislice cross talk	7-5
	7.2.4 Dielectric effects	7-6
	7.2.5 Aliasing	7-8
7.3	Advanced ideas that are more concerned with frequency space	7-9
	7.3.1 Truncation/Gibbs	7-9
	7.3.2 Zippers	7-10

	7.3.3 Corduroy artifact (spikes)	7-12
	7.3.4 Moiré fringes	7-13
	7.3.5 Ghosting	7-14
7.4	Material control	7-17
	7.4.1 Chemical shift (Type 1)	7-17
	7.4.2 Chemical shift (Type 2)	7-18
	7.4.3 Magic angle	7-20
	References	7-24

8 Concluding a journey through MRI   8-1

Preface

We have formed this book that uses a conceptual approach as one that has links and a format that is a little different from others, as it includes animations and videos, encourages other sources, and hopefully connects some of the dots with topics about which you may have been interested in learning more. Perhaps it can be used as a bridge from the different approaches ground-up, from proton physics to webpage and to outline for board preparation and to applications. Just as Dr Elster comments, 'there still exists common confusion that has not been answered satisfactorily by existing books and web sources.' We hope to reward the patient reader by enhancing critical thinking skills and to further create sense and meaning from the vast and complex MRI concepts.

First of all, I want to thank the many colleagues and students who have improved this work: Marlee Reust (senior lab analyst and an amazing illustrator); Caroline (Carlie) Preskitt (dancer and devoted healthcare professional); Natalie Norton (Pre-med Biomedical Engineer, and former athlete); Gail Harmata, Ph.D. (thoughtful neuroscience post-doc); Kwong Hui, Ph.D. (eternal classmate/friend), Whitney Geis R.T. (R)(MR) ARRT (MR Safety Officer); Justin North, M.D. (Radiology Attending); William Sattin, Ph.D. (MR physicist); Amy Barnett; Nicholas Psencik (MS3); Cole Lindley (MS4); Willow Arana; Nathan Vo; Phillip Rhoton; Daniel Lawerence (R)(MR) ARRT, Asa Brown; Kaustav Sahoo Ph.D.; Bennett Hogan M.S.; Everett Cavanaugh M.S.; Monica Senese; Kathy Kyler; Trista Mythen; Eng-u Wu (my supportive brother); and of course my loving wife, Hannah B Swallow, Ph.D. and three wonderful children Ian, Oliver, and Lillian, who make all things possible. Finally, I want to thank the IOP Publishing team for assisting me in establishing a new way to present information using their e-book platform.

Finally, while the field of MRI is vast and can be daunting for the beginner, it is not impossible to grasp. Whether you are a radiology resident and/or a technologist, or any practicing physician attending, we hope these materials enable you to better connect the dots in MRI. Additionally, medical physicists, researchers (such as in psychology/neuroscience), and/or biomedical engineers who are interfacing with medicine can also get a window on both technology and clinical concepts. We start with five core principles in chapter 1, then five hardware components in chapter 2. Next, in chapter 3, we discuss relaxation essential for contrast and resolution in MRI. Chapter 4 dissects the initially mysterious pulse sequence diagram that I have been asked to cover many times by physicians, radiological technologists, and medical physicists. In chapter 5, we included a few topics to bridge a couple of 'missing' concepts that include transmit and receiver bandwidth and coverage of k-space for resolution and contrast. Chapter 5 is meant to aid the reader in bridging to the final important and key clinically relevant ideas of the closing chapters 6 and 7. The topics that we hope to emphasize include chapter 6 on pulse sequence tradeoffs in time, signal-to-noise ratio, and heating, and chapter 7 on artifacts recognition, sources, and potential remedies. All of these are important components, as they drive the efficiency, reliability, and performance of a well-functioning MRI center, and are topics that are currently a focus for board evaluation. Driven by our desire to understand and look within the body at its structure, function, and anatomy, MRI is an exquisite modality that is a great science with an ultimate utility that serves our patients. It is a pleasure to try to express my passion for it in this format, and I hope you find this a useful tool to augment your own understanding.

Dee H Wu, MSSE, Ph.D., DABMP, DABMRS
University of Oklahoma Health Sciences Center

Acknowledgement

For my parents Ta-Sun and Ling-Tung Wu.

Author biography

Dee Wu

Dee Wu, Ph.D. currently serves as Professor and the Chief of Technology Applications and Translational Research in the department of Radiological Sciences at the University of Oklahoma Health Sciences Center (OUHSC), where he has been since 2002. Dr Wu started his MRI career at the University Hospitals of Cleveland over 30 years ago as a programming analyst and research assistant. He received his doctorate degree under Dr Jeffery Duerk and Dr Jonathan Lewin. He was previously a clinical scientist with Picker/Marconi Medical Systems and then briefly with Philips Medical Systems where he worked on designing and constructing applications in industry before joining the medical physics and radiological science team at OUHSC.

Dr Wu has devoted his career to better understanding imaging, particularly in the hospital environment. His experience has spanned morphological, biophysical, neuro-functional, and metabolic alterations associated with disease. Dr Wu has been a faculty member of the Oklahoma Center for Neuroscience (OCNS) and the Department of Geriatrics, and has been an associate member of the Stephenson Cancer Center. Further, Dr Wu is a board-certified diplomate in Magnetic Resonance Imaging Physics from the American Board of Medical Physics (DABMP) and has over 30 years of professional experience working both in industry and academic medicine.

Dr Wu has 14 patents regarding the development of MRI techniques and applications, and 55 peer-reviewed publications in relevant MRI-related journals. Additionally, Dr Wu's technical skills have led to affiliate/adjunct faculty member status in both the Department of Computer Science at the University of Oklahoma (Norman) and the department of Electrical Engineering at the University of Oklahoma (Norman). Dr Wu is currently the chair of the American Association of Medical Physicists in Medicine Medical Physics 3.0 Smart Expansion committee. Dr Wu has had translational projects in far-reaching subspecialty areas of medicine (from neonatology, neurosurgery, genetics, oncology, emergency medicine to geriatrics, etc). His passions are aimed at practical implementations that seek to improve medicine and society by novel technology use.

Dr Wu further relies on broad but solid physics/computational/engineering skills that are then applied to real-time problems and challenges that occur in medical practice. He has worked on Bruker System's sequences specifically for development work on pre-clinical MRI scanners to collaborate directly with molecular biologists. He complements this with his experience in pulse programming experience on GE, Siemens, and Philips scanners that he has developed over the span of his 30-year career. Finally, at OUHSC, Dr Wu heads the Mentorship and Scholarship of the Academy of Teaching Scholars, which has over 20 medical faculty members, as a co-chair for that group. Dr Wu has mentored multiple students at various stages of degree completion and professional development, and is a passionate teacher and educator.

How to use this book

Welcome to *MRI—Connecting the Dots: A conceptual approach*. This e-book is intentionally designed to be read by different types of learners. Whether you are a medical resident, technologist, radiologist, MRI scientist, or anyone else curious about MRI, there may be different paths to how readers might like to receive this information.

Where to start? The answer is where you, as the reader, feel most comfortable. The following matrix is provided to suggest where the reader could start in this book based on their current background as seen in figure A1.

* = first choice for starting reading
X = second choice for starting reading
O = third possibility for starting reading
• = later topic for reading.

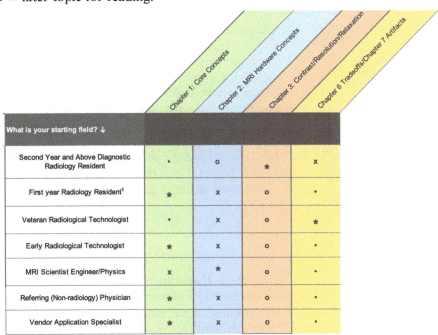

Figure A1. Matrix to determine where the reader should begin based on their previous experience[1].

---

[1] Some veteran radiological technologists have read through chapter 6, which concerns signal-to-noise ratios (SNR) and time tradeoffs in pulse sequences. Then, they looked through chapter 7 on imaging artifact review before returning to earlier chapters that support understanding these concepts (chapters 1–5). Additionally, the goal of medical physicists and MRI scientists is to be able to discuss and explain the concepts in chapter 6 and chapter 7 through the fundamental understanding of mechanisms (i.e., topics in chapters 1–5). However, this communication should be done in a way that appears 'applicable' to radiologists and radiological technologists. Also, communication by the physicists and scientists should be done in collaboration, as there is still much to learn about the patient-facing side of the service that can be mutually gained by the communication with their fellow healthcare workers.

I Where and how to start reading?

Chapter 1 provides five core concepts upon which the science of MRI is built. Chapter 2 outlines the five pieces of hardware that together make the MRI scanner. Think of the fundamental nature of these concepts as the fingers on your hands: five core principles, five fingers on each hand. You can build strength and teamwork skills by developing both hands, or simply start with one hand (chapter 1 or chapter 2). In the lectures I give, I cover both chapters in a single lecture, and provide handouts from my book for more inquisitive first-year residents. For an allied health, medical physics, or biomedical engineering class, I present two lectures and provide additional resources over more basic, science-oriented materials. So, based on the reader's background and interests, there are three possible places to begin reading this book. The reader should choose one of three places to start:

Chapter 1: **Five big ideas for MRI**
For the learners that like to start with fundamentals, start at chapter 1 and proceed onward. Chapter 2 is important, as it represents a concrete explanation of MRI (i.e., the hardware/equipment). Understanding MRI hardware is crucial for communicating with technologists, physicians, and scientists. We must also consider how the hardware operates from a safety point of view. Chapter 1 can be helpful for preparing to understand artifacts, a subject that is covered in chapter 7.

Chapter 2: **Hardware**
If you are an MRI scientist, allied health or medical student, and/or technologist, then I recommend that you start with the hardware components and skim and/or refer back to chapter 1. Core concepts of chapter 1 can help the reader connect fundamental physical science topics with medical concepts. It is useful to know these ideas if the reader interacts with the clinical environment and/or hopes to conceptualize deeper MRI topics, but starting with hardware may be a good place to start your journey into MRI.

Chapter 3: **Basic building blocks**
If you are a resident and already integrated into clinical environments, you may choose to begin with content that captures the 'big picture' rather than building upon concepts[2]. We encourage those with a sufficient background in MRI to jump to later chapters in this book while referring back to chapter 1 and chapter 2 as needed. Chapter 3 dives into describing signal contrast and resolution. Some of the strongest clinicians, chair-holders, and leaders in the MRI field grasp deeper concepts and continuously work to improve their craft so they can adapt to new technologies and to 'connect the dots' through concepts.

No matter where the reader is coming from, the content in this book is provided to introduce the core principles in a logical fashion to help all learners. Regardless of background, I highly recommend looking at the interactive components, particularly on waves. MRI is built on waves, and if you better understand waves, it may help you grasp some of the other concepts more readily. This is useful whether you are looking backward in approach by reading later chapters and then coming back to earlier core ideas, or if you are developing a forward approach by starting with the fundamentals in this first chapter, as depicted in figure A2.

II How to use the 'FLASHY' box system

The book utilizes text 'bubbles' that highlight non-crucial information. The text bubbles are categorized into six types: Fun-fact, Literature, Applied, Scaffolding, Historical, and Yield. Shorthand, we refer to these as 'FLASHY' bubbles. The book is meant to be read without pausing to read over the FLASHY 'bubble'. These 'bubbles' of topics are used to either preview or refine knowledge and permit the curious reader to reflect more. However, the book is also written to permit the reader to move more quickly by skipping the bubbles to concentrate on essential ideas if time is limited. The reader will ultimately be led to two prime interests: (1) making tradeoffs in MRI between signal-to-noise, time, and controlling energy deposition; and (2) MRI image artifacts. These two topics seem to engage the interest of radiology residents and technologists.

---

[2] Understandably so, time may be limited during residency and primarily focused on clinical learning. It would be most useful to begin with the chapters most relevant to your current mindset. The chapters are written to easily reference earlier fundamental concepts should further clarification be needed.

Figure A2. The FLASHY[3] 'boxes' system connects the reader with one of the thematic approaches spelled out in the acronym letters. They are meant to stand alone; the reader who may be rushed to get through the material can skim these 'boxes' without losing much in their conceptual understanding. Adobe Stock Images: © N.Savranska.

- **'Fun-fact' bubble**—meant to stimulate interest. It may have a fun-fact or emotional resonance that helps you to remember concepts. It is not fully meant to be instructive, but this boxed material can add whimsy and fun to the learning process.
- **'Literature that is evolving' bubble**—connects you with evolving literature.
- **'Applied' bubble**—a small summarizing snippet of information on how you can apply the knowledge or concepts to solve a clinical problem or real-life scenario.
- **'Scaffolding' bubble**—lays the foundation for something covered later on, and will help you understand concepts. It differs from the applied bubble, as it may be more abstract or intangible at the moment. You may only see secondary effects of what it does. Nevertheless, these will be important to your understanding.
- **'Historical-reference' bubble**—connects you with old references that were instrumental in the development of MRI concepts and technologies. This could be an 'oldie but goody' piece so you can check sources and look at how past scientists, physicians, and other people figured things out.
- **'Yield (significant yield) question' bubble**—slightly 'worked' through provocative questions that require some understanding of the materials. We believe these are topics that may be more frequently evaluated on tests for conceptual understanding.

Wherever you choose to start in this book as a radiology resident, physician, more seasoned technologist, or MRI scientist in training, I hope you find the 'FLASHY' bubbles and Deductive Learning Sketches to be useful. A Deductive Learning Sketches (DLS) jumps to a real world example, asking the reader to independently work through the question by learning to apply what they have learned. We will then discuss the example in more detail. In addition to the bubbles, we intersperse real readers who are practicing MR technologists, physicians, medical students,

---

[3] FLASHY is an acronym for Fun-fact, Literature, Applied, Scaffolding, Historical, Yield, and refers to the 'boxes' system used in this book to separate thematic topics that are adjunctive to the main content provided in the 'unboxed' sections.

psychology and neuroscience researchers, and premedical students who reflect on sections to provide authentic reactions and voices to help the reader connect with not only the technological content, but also the clinical content. In particular, for educators of residents or technology students, we like to focus on the 'key' fundamental ideas and provide handouts from the book to supplement a deeper understanding of the topic. A goal would be to eventually lead the learner to engage with and understand resources developed for clinicians, such as MRIQuestions.com by Allen Elster, MD and Radiopedia™, and for those with specific interests to watch relevant topics of interest to them from Michael Lipton M.D., Ph.D.'s 56 YouTube online MRI lectures series.

For radiological technologists, the choice could be based on where they are in their career: some veterans can also follow a plan that starts with advanced chapters (chapter 2 or chapter 3) and then return to fundamentals at a later time, and in this way connect and augment their day-to-day practice. For example, I would consider this a form of looking forward to the themes that you might encounter in real life, like those in chapter 6 or chapter 7, and then going back to learn fundamentals when/if needed to strengthen understanding. For radiological technologists early in their careers, we highly recommend that these readers proceed with these chapters in order. In this way, they can create a learning scaffold aimed at connecting the learner to the application of MRI and to enable them to better adapt to new and emerging concepts as the technology evolves.

For those who seek to be MRI clinical scientists (medical physicists, biomedical engineers, and others), these chapters can provide insights into how clinicians relate to the technology. It is highly recommended to read the chapters in order and then move forward to guide their knowledge building. While this book broadly aims at clinical learners and develops core concepts without heavy mathematics, a deeper and more fundamental understanding will be achieved by being able to read through classic references, such as that from *MRI Imaging* by Brown *et al* and *MRI Handbook of Pulse Sequences* by Bernstein *et al*. These resources include a mathematical, physics, and engineering approach to learning MRI that is adjunctive to starting with the themes and principles of this book.

Some referring physicians and/or even allied science professionals interested in the topic will hopefully find that these concepts provide a missing link that will help them build knowledge. For these learners, I highly recommend looking at the interactive components (such as animations and/or interactive online sources). Even if you don't easily fall into one of the above professional categories (perhaps you are a referring physician or allied science professional and/or casual reader interested in the topic), I still hope that you find these concepts practical. I have used parts of the content to instruct non-radiology attendings and allied health professionals, such as experimental psychologists, medical students, and neuroscience students over many years, and asked for comments from a variety of people interested in MRI.

Throughout the book, I have interspersed reflections by readers who are medical students, premedical students, residents, technologists, attendings, and scientists to help the reader potentially 'dialogue' with the concepts in this book. Here are the 'cast of characters' who have with great appreciation for their thoughtfulness and insights propelled me throughout the writing of this book.

Finally, regardless of which domain you as the reader may be coming from, the content of this book is provided to introduce the core principles in a logical fashion to help all learners. We hope that this approach to content is useful whether you are looking backward by starting with reading later chapters and then coming back to earlier core ideas, or if you are simply developing a forward approach that starts with the fundamentals in the first chapter.

Introduction for residents, technologists, scientists, and for anyone curious

Magnetic Resonance Imaging (MRI) is a discipline encompassing a vast, challenging, and fascinating amount of knowledge that crosses over numerous domains of physical and medical sciences. There are many ways that people come to be involved with MRI, from those engineers and service professionals that build/augment the inner workings of the scanners, technologists that diligently operate the scanner, those that research new uses, and the many physicians that use the images diagnostically. This technology has captivated many, and a countless number of people have contributed to new scientific discoveries and applied their skills and knowledge to impact a countless number of lives.

MRI is useful to people from many different backgrounds and application areas, including:

a. Technologists who have some experience on the scanner, but also may like to receive a deeper familiarity with MRI concepts;
b. Residents/Fellows who are curious and want to have a 'scaffold' of knowledge to understand what is happening as they interact with the hospital's imaging world;
c. Allied professionals, such as neuroscientists, psychologists, and/or post-docs in other fields, who may be interested in understanding MRI when they encounter it directly or peripherally;
d. Medical physicists or biomedical engineers who need a brief introduction; and perhaps they could be looking for materials that could be useful for them teaching others, perhaps using the material as handouts for radiology lectures they may provide for other disciplines[4];
e. A casual learner, such as a vendor, preferably with a little bit of engineering and math (STEM) background, who wants to gather more clinical-specific knowledge on how MRI works from a conceptual point of view.

The primary goal of this book is to provide an informative and initial summary of key MRI principles for the broad ranges of individuals who work with MRI machinery, but come potentially from a multitude of backgrounds. However, in any craft, it is still important to have some practice with these science fundamentals to gain a more robust orientation to the content area of MRI. Pre-medical students and radiological technology students are already required to take several levels of

---

[4] Additionally, therapy medical physicists who are interested in MRI-RT could find the sections/chapters on MRI Pulse Sequence Tradeoffs and MRI Artifacts to be a quick start into understanding concepts around problem solving clinical issues for image quality and machine calibration purposes.

chemistry and some physics. So, we start with some basic assumptions with which the reader might have a little bit of prior exposure. Luckily, for those whose fundamentals are relegated to the past, the internet is teeming with resources for reviewing and beginning to learn these STEM topics, and there are many techniques for eventually gaining mastery. Leveraging strong content from Chem-libretext, Hyperphysics, EdX, Coursera, and Khan Academy websites, readers can bolster their fundamental science understanding. Alternatively, readers can find the contents they want for review just in case they have forgotten parts of the earlier fundamental materials to which they were exposed.

A basic chemistry and/or physics class will support the reader's understanding of MRI, as well as medicine. We've had radiology residents, medical students, pre-med students, radiological technologists, medical physics, undergraduate computer and electrical engineering, and psychology students able to grasp the contents of this book (enough to gain new insights), and it even encouraged some to take a chemistry and physics course to connect their knowledge with higher levels of detail. In particular, prior exposure to electric and magnetic charge, the Bohr model, dipoles, force, torque, waves, and energy can be helpful. We hope that the growth in knowledge about MRI may lead the reader back to some fundamental concepts as they continue and as their passion leads them in this topic.

We hope this book is useful for those who are curious to learn more about MRI, but also for those who need an adjunctive study aide. Such tools need to change continuously, and we may update frequently. However, conceptual ideas remain more consistent for a longer period of time. We realize that board review and exams may bring the readers to seek more knowledge. But, we also wish for clinicians and scientists to stay true to the initial passion and curiosity with which they first came to the field, and hopefully refresh their passion for the sciences.

I currently serve as the Chief of Technology Applications Development and Translational Research in the Department of Radiological Sciences at the University of Oklahoma Health Sciences Center; Adjunct/Affiliate Associate Professor in the School of Computer Sciences; and Adjunct/Affiliate Associate Professor in the School of Electrical Engineering at the University of Oklahoma. The value of communication among the clinical sciences was stressed in my training and reinforced by my years as a clinical scientist at Philips Medical Systems Cleveland and Marconi Medical Systems. I am a DABMP board-certified medical physicist and Diplomate of the American Board of Magnetic Resonance Safety (DABMRS).

I was lucky to be mentored by several fellows of the International Society in Magnetic Resonance and Medicine (ISMRM) and pioneers of their field, including Mark E Haacke and Jeffrey Duerk. I received my Ph.D. with two advisors, Dr Duerk, a medical physicist, and Dr Jonathan Lewin. Dr Duerk is currently the Executive Vice President for Academic Affairs and Provost at the University of Miami. Dr Lewin is a physician who had served as chair of the University Hospitals of Cleveland, then Johns Hopkins Radiology, and now is the Executive Vice President for Health Affairs at Emory University. Also, formative were mentorship and encouragement by many accomplished industrial clinical scientist colleagues early in my career, including Wayne Dannels, David Foxall, and James Murdoch.

Further, I have been amazed by the commitment of many radiologists, including my current mentor, Chair of Radiological Sciences Anthony Alleman, M.D., who also was/is an Otolaryngology Surgeon and Public Health specialist. From the teaching and leadership side, I want to acknowledge my many colleagues in the Academy of Teaching Scholars (ATS), where I currently serve as Mentoring and Scholarship co-chair. I believe that teaching is an organized activity that nurtures learning at all levels of education, and it is one that can be enhanced when shared between different disciplines. As subspecialties become increasingly more specialized, a greater need to provide opportunities for dialogue and interchange emerges. I'm hoping to help colleagues use tools to develop their education and service skills.

So where did this book come from? As a medical physicist, applications scientist, affiliated healthcare specialist, and bioinformatician, I have come to value the full range of professionals with whom I collaborate in the hospital with a goal of improving healthcare. The application of technology requires being able to converse and work with a multitude of disciplines. OU Medicine, our hospital system, has over 11,000 employees and over 1,300 physicians and advanced practice providers. Over the years, I have had the privilege of working alongside numerous physicians and practice providers across a wide variety of medical disciplines. This work has spanned subspecialties from neonatology to geriatrics and from medical genetics to neurosurgery. As such, I currently serve as a committee member on the Medical Physics 3.0 initiative for the American Association of Physicists in Medicine (AAPM). This committee seeks to enhance the full value of physics towards human health, including clinical practice, administrative, scientific, and educational opportunities, as well as to seek and identify and explore other areas where medical physicists can assist in healthcare. Each day, I hope to educate colleagues and learners about how to use their equipment more effectively, to develop a stronger safety culture, and to gain greater knowledge of the principles they are 'applying.'

There are so many sources of information on MRI on the internet. I can only list a few from which I have grown to discover more and more ideas and information. Breaking these down into domains that include the interface of technology and the clinical world of MRI, MRIQuestions.com stands out as a personal favorite. In fact, if I can get Radiologists, MR Technologists and/or Medical Physicist Residents and students to have the aptitude to read this site, I would consider it a big teaching/mentorship win. This is maintained and written by Alan Elster, M.D., from Mallinckrodt Department of Radiology and previous chair of the Wake Forest Radiology department. This content, according to Dr Elster, presents concepts in a way that is radically different from 'standard' Q&A books. These were not questions of his own creation, but were questions actually posed to him by trainees over the years. As such, they reflected areas of common confusion that had not been answered satisfactorily by existing books and web sources. The other clinical asset that I use is Radiopedia [1], which provides content open for editing articles and allows radiologists, radiology residents/radiographers, and other healthcare professionals to modify and refine the content and updates. I hope that the content of our book will be an entree to the aforementioned clinically bridging websites as the reader grows their interest over time.

If you are a reader who wants to increase your understanding along a physics- or an engineering-based trajectory, I highly recommend the book *MRI Imaging* by Brown *et al*. I use this resource mainly with our medical physics and engineering graduate students. This book stands as one of the most important educational books in the field of MRI. The Brown *et al* book can be well complemented by the *MR Handbook of Pulse Sequences* by Matt Berstein *et al*, which is a thorough and thoughtful collection of highly relevant 'recipes' that dives deep into the construction of pulse sequences applicable for both engineers and scientists. Much of this knowledge in Bernstein's book is not readily available outside the confines of the industrial laboratories that construct these machines. I also want to thank Dr Larry Wald, Mass General Hospital (MGH), for the hands-on experience he provided me in understanding more about MRI coil development. This helped me bridge some of my gaps in the art of coil-making that I was able to gain on a brief visit to him. A good starting book concerning MRI coil building is *RF Coils for MRI* by Thomas Vaughan, with advanced details on the subject of RF coil construction [2].

Sample pathways for learning for different professionals
You may choose to use this book as teaching materials to provide additional structure to your lessons and to assist in your delivery of instruction. As illustrated in figure A1, there are many potential places to start in this book, based on the reader's background. We acknowledge there are many sources of information on MRI throughout the internet and other books, and we encourage the reader to explore these resources to aid in their understanding. The book serves as a potential way to connect some of these resources for the learner, which can help improve their understanding and fluency. Below are a few suggestions on pathways for learning that have worked for the author.

For radiologist residents
During the first year (three lectures), we cover concepts in the first three chapters. Chapter 1 contains five fundamental biological and physics concepts. Chapter 2 concerns the five components of MRI hardware scanners. Upon learning ten concepts, in chapter 3 we cover relaxation and spin physics, which is the cornerstone of MRI contrast and physiological differentiation of disease. The ideas in chapter 3 set up the future topics in the second-year lectures.

During the second year (seven lectures), we start with a review of relaxation in chapter 3, since it has been one year since they explored the concept and it is central to understanding later ideas. Chapter 4 provides the fundamental concept of pulse sequences, so that the concept is not foreign and becomes less mysterious. Chapter 5 has three essential ideas that are related to pulse sequences: transmit bandwidth, receiver bandwidth, and how pulse sequences cover k-space. In this second year, we also include additional advanced topics that are not covered in this introductory book, but are relevant for board preparation. These topics include inversion recovery, diffusion, cardiac, breast imaging, angiography, and so on [3–6].

During the third year (three review sessions), we focus on board review questions. This approach is based on the RSNA review modules, Imaging Physics Case Review Book, and Radiologic Physics War Machine Book [7–9].

For veteran radiological technologists
These professionals may consider reading chapter 6 on MRI tradeoffs and chapter 7 on MRI artifacts first. Both chapters are pragmatic and used in the clinic on a day-to-day basis. Further, these veteran technologists can review chapters 1–5 to strengthen their foundations and to further connect the dots of their understanding, as needed.

For beginning technologists and those preparing for boards
Beginning technologists can rely on primary text, such as *MRI in Practice* by Westbrook and Talbot [10] or *MRI Physics: Tech to Tech Explanations* by Powers [11]. Then, they may focus on chapter 6 on MRI Tradeoffs and chapter 7 MRI Artifacts for board preparation and use chapters 2–5 to aid in their understanding of hardware, relaxation, and pulse sequences.

For veteran radiologists
Start anywhere in the book that you are curious about. You might use the book to springboard your ability to read MRIQuestions.com.

For vendors who are application specialists
With a little bit of chemistry and physics, this book could help you better understand what happens in the radiology environments. Vendors who conduct technical sales or account executives would find some of the materials, such as our animations, could help them better connect with the radiology team.

For medical students and referring physicians
It may be helpful to develop knowledge of concepts and vocabulary so that you can better connect with your radiology colleagues. In this way, you can guide your patients to achieve the best results and build your practice.

For master's level medical physicists/residents
You can use this book to review medical physics 3.0 practices, which include being able to communicate effectively with fellow healthcare professionals. The book can be used to supplement the MRI chapters in *The Essential Physics of Medical Imaging* by Bushberg *et al*, *MRI from Picture to Proton* by McRobbie et al [12, 13]. Medical physics readers can use the diagrams and additional explanations to break down the materials to greater generalization. Based on feedback in 2021 by our medical physics residents, they felt that the book would set grad students up for solid preparation on the MRI portion of the ABR part one exam.

For MRI scientists/engineers
Use this book as a springboard to understand the role of clinical scientists and/or better conduct research and development. The book can also help you start your preparation for boards in the MRI Physics Specialization.

For practicing/veteran medical physicists
Use the book to supplement other sources, such as Brown *et al* [14], Bernstein *et al* [15], and Nishimura [16], which contain a more mathematical treatment of the topic.

This book be used to facilitate communication with radiologists and technologists. Additionally, this book can be used to consider eventual board certification in MRI and/or MRI-RT.

Finally, it would be impossible to cover all of the books and resources from which one can learn about MRI. I would highly encourage the reader to find which content sources match the reader's interests and stimulate their curiosity. If I had to choose just five resources (a hard choice) that I would use to draw ideas to teach from, beyond those previously mentioned, I would consider the following:

    a. *MRI made easy: (...well almost)* by Hans Schild—This resource includes some wit and whimsy, and it uses a cartoon illustrative style. It is a good starter for relaxation and the spin echo, which has also been reported to me by learners. It is out of print, but some modified versions are available online [17].

    b. *Crack the Core* by Promeus Lionhart is a resource starting point for diagnostic radiology residents. It is a great resource for the busy mindset of residents, as it provides focused materials and simple outlines. It is updated and clear with the bare minimum that you need to know for the diagnostic radiology boards in terms of physics [9].

    c. *MRI in Practice* by Catherine Westbrook This book is a useful starting point for radiology technologists. Our radiology technologists in training quote this book many times [10].

    d. Xrayphysics.com by Mark Hammer, M.D., who created it when he was a medical resident. It illustrates concretely many of the examples, including some simulations [18]. Well done!

    e. *Introducing MRI Videos* by Michael Lipton is a tome of 56 free lectures online that assumes non-technical and practical information [19]. It is inspiring to me to see a physician demonstrate such passion and interest in physics. It is even a reminder for myself to take an equivalent passion in understanding medical school topics, which for myself include immunology, genetics, molecular biology, pathology, physiology, and anatomy. I strive to understand these topics on a day-to-day basis to be a better provider of my craft. Our residents enjoy Lipton's resource, especially as they can pick and choose the relevant content from it.

There are also other quality references that I have learned about from different students and colleagues. I browse these resources from time to time, especially in terms of inspiration for teaching ideas. I'm listing ten more from those I know about below.

    1. Donald McRobbie's *MRI Picture to Proton*—Dr McRobbie is a medical physicist from the UK. The book has excellent figures and engaging content that is especially useful for early medical physics students as a starter [13].

    2. Perry Sprawls' *Magnetic Resonance Imaging: Principles, Methods, and Techniques* has great PowerPoint slides with appealing style and a high level of content. It is a source for clarifying much of the knowledge not only of MRI, but also of other medical physics areas [20].

    3. Kiaran McGee's *Mayo Clinic Guide to Cardiac Magnetic Resonance Imaging* (Mayo Clinic Scientific Press), 2nd Edition. I've enjoyed this resource, especially as it focuses on Cardiac MRI [21].

4. Ray H Hashemi, M.D., Ph.D.'s **MRI: The Basics**. I really like this book as it includes a physician's view on physics. Particularly enlightening are the descriptions for some of the applications, such as angiography and cardiac MRI [22].
5. Joseph P Hornak's **The Basics of MRI** website—Dr Hornak was one of the first Ph.D.s to put materials on the internet that was shared freely. This has helped many learners since 1996, and is still being updated to this day [23].
6. Dwight Nishimura's **Principles of Magnetic Resonance Imaging**. Nishimura's book presents the basic principles of magnetic resonance imaging (MRI). The emphasis is on the signal processing elements of MRI, particularly the Fourier transform relationships, as it was developed as a teaching text for an electrical engineering course at Stanford University [16].
7. Michael Lipton, *Totally Accessible MRI: A User's Guide to Principles, Technology, and Applications*. This book could be nice adjunctive material to accompany his YouTube lectures that I previously mentioned [24].
8. **Brian Dale** updated and edited one of the early books entitled **MRI: Basic Principles and Application,** which was first created by Mark Brown in 1995. It still is an enjoyable entree to MRI and includes very valuable insights from Richard Semelka, M.D., of body MRI publishing and knowledge fame [25].
9. Stewart Bushing and Geoff Clarke's *Magnetic Resonance Imaging: Physical and Biological Principles*. I personally met both medical physicists and have enjoyed talking to them. They are both very committed to the field of MRI and their book is timeless, with great illustrations and physics descriptions [26].
10. Jerrold T Bushberg's ***The Essential Physics of Medical Imaging***. This book has been a staple in terms of a summary of the entirety of Medical Physics, and has a couple of introductory chapters on MRI. Medical Physics residents study heavily from this source, which is not only for MRI, but also contains other topics [12].

All of the many sources above are great, have received ample interest, and provide different ways of explaining physics. There have been many more over the last 30 or so years, and I wish I could list them all.

Dee H Wu, MSEE, Ph.D., DABMP, DABMRS
University of Oklahoma Health Sciences Center

References

[1] *Radiopaedia.org, the wiki-based collaborative Radiology resource* (n.d.) Radiopaedia. Retrieved April 3, 2022, from https://radiopaedia.org/
[2] Vaughan J T and Griffiths J R (eds) 2012 *RF Coils for MRI* (1st edn) (New York: Wiley)
[3] *T2-FLAIR* (n.d.) Questions and Answers in MRI. Retrieved August 16, 2022, from http://mriquestions.com/t2-flair.html
[4] *Diffusion* (n.d.) Questions and Answers in MRI. Retrieved August 16, 2022, from http://mriquestions.com/diffusion-basic.html
[5] *Breast DCE* (n.d.) Questions and Answers in MRI. Retrieved August 16, 2022, from http://mriquestions.com/breast-dce.html

[6] *MRA methods* (n.d.) Questions and Answers in MRI. Retrieved August 16, 2022, from http://mriquestions.com/mra-methods.html

[7] *Physics modules* (n.d.) Retrieved August 16, 2022, from https://www.rsna.org/education/trainee-resources/physics-modules

[8] Abrahams R B, Huda W and Sensakovic W F 2019 *Imaging Physics Case Review* (New York: Elsevier)

[9] Lionheart P 2017 *Radiologic Physics—War Machine* (independently published)

[10] Westbrook C and Talbot J 2018 *MRI in Practice* (5th edn) (New York: Wiley–Blackwell)

[11] Powers S J 2021 *MRI Physics: Tech to Tech Explanations* (New York: Wiley)

[12] Bushberg J T, Seibert J A Leidholt E M Jr and Boone J M 2011 *The Essential Physics of Medical Imaging* (3rd edn) (Philadelphia, PA: Lippincott/Williams & Wilkins)

[13] McRobbie D W, Moore E A, Graves M J and Prince M R 2017 *MRI from Picture to Proton* (Cambridge: Cambridge University Press)

[14] Brown R W, Cheng Y-C N, Haacke E M, Thompson M R and Venkatesan R 2014 *Magnetic Resonance Imaging: Physical Principles and Sequence Design* 2nd edn (New York: Wiley)

[15] Bernstein M A, King M F and Zhou X J 2004 *Handbook of MRI Pulse Sequences* (Oxford: Elsevier)

[16] Nishimura D G 1996 *Principes of Magnetic Resonance Imaging* (Stanford, CA: Stanford University)

[17] Schild H H 1997 *MRI: Made Easy* (2nd edn) (Berlin: Schering AG)

[18] XRayPhysics—Interactive Radiology Physics (n.d.) Retrieved 3 April 2022 from http://xrayphysics.com/

[19] Albert Einstein College of Medicine 2014 *Introducing MRI: The Basics (1 of 56)* https://www.youtube.com/watch?v=35gfOtjRcic

[20] Magnetic Resonance Imaging (n.d.) Retrieved April 3, 2022, from http://www.sprawls.org/mripmt/

[21] McGee K, Williamson E and Martinez M (eds) 2015 *Mayo Clinic Guide to Cardiac Magnetic Resonance Imaging* (2nd edn) (Oxford: Oxford University Press)

[22] Hashemi R H 1997 *MRI: The Basics* (Philadelphia, PA: Lippincott/Williams & Wilkins)

[23] The Basics of MRI (n.d.) Retrieved April 3, 2022, from https://www.cis.rit.edu/htbooks/mri/inside.htm

[24] Lipton M R 2010 *Totally Accessible MRI: A User's Guide to Principles, Technology, and Applications* (Cham: Springer)

[25] Dale B M, Brown M A and Semelka R C 2015 *MRI: Basic Principles and Applications* (5th edn) (New York: Wiley–Blackwell)

[26] Bushing S C and Clarke G 2013 *Magnetic Resonance Imaging: Physical and Biological Principles* (Amsterdam: Elsevier)

IOP Publishing

MRI: Connecting the Dots
A start to concepts

Dee Wu

Chapter 1

Five big ideas for MRI

(Waves, water, multiple looks, dipoles/precession, and changing EM fields)

1 Introduction

About four years ago, after showing a neurosurgeon a list of the last 50 Nobel Prizes in Medicine, I asked that neurosurgeon which of those innovations he considered to be the most impactful in his daily work [1]. He took some time to think over the list, but, in the end, the neurosurgeon put Computed Tomography (CT) and Magnetic Resonance Imaging (MRI) at the top of the list. These two technologies guide diagnosis of conditions, identify the locations of lesions (tumors, aneurysms, etc), and even direct therapies, such as Laser Interstitial Thermal Therapy (LITT), among other treatment possibilities. This is not to say that any of the other Nobel Prize findings were of a lesser magnitude[1], but, in his daily practice, the innovations enabled by MRI and, more generally, imaging technology could not be ignored. Beyond neurosurgery, CT and MRI impact the development and application of many clinical disciplines.

Almost every department in a hospital (including Emergency Medicine, Neurology, Neurosurgery, Pulmonology, Hematology–Oncology, and almost all clinical subspecialties) will refer[2] cases to the Radiology Department for MRI daily. Scaffolding concepts can be useful for improving understanding and communication in the hospital across these departments. It is thus important to understand the five concepts highlighted in this chapter because medicine is built on teamwork. We can infer that MRI is certainly a powerful diagnostic tool in medicine just from the number of exams and diagnostic consultations that are requested and conducted each year. The University of Oklahoma Health Sciences Center (OUHSC) and neighboring

---

[1] Some disciplines, such as Hematology–Oncology, may rate other Nobel prizes higher from the list taken from the last 50 years. We do not mean to diminish any other accomplishments, especially in Nobel prizes, but we hope to advocate for what MRI can do for medicine from different subspecialists' points of view

[2] In radiology, we refer to all physicians who are not radiologists as referrings. The 'referrings' are the ones who are seeking answers to medical problems, which are focused on patient care.

Veterans Administration Health Care System (VAHCS) have over 10 imaging systems that run almost around the clock to meet the demand of imaging services.

MRI not only provides obvious and critical clinical utility, but it also offers advanced research opportunities in the study of pathophysiological processes [2]. MRI provides one of the most exciting glimpses into the human body, physiology, and even the mind for researchers in Psychology, Neuroscience, and Rehabilitation Sciences[3]. As of January 2022, a search on PubMed for MRI has returned 685,510 search results.

Learning about MRI is a journey that many physicians, radiological technologists, and researchers begin in their training each year. In this book, we hope that we approach this topic in a way that increases your ability to consider the elegance and importance of MRI in both a mechanistic and a clinically applicable way. We will guide you in the core concepts that bring you to the important and influential topics of making **tradeoffs in MRI** pulse sequences that we cover in chapter 6, and another vital topic, **MRI artifacts**, that we cover in chapter 7. We will also develop your conceptual understanding by directing you to resources that may help you on this journey. There are many educational mediums, including YouTube videos, other books, blogs, and even courses, that you could use to consolidate your learning.

What we are going to do now is get you started by providing you with the initial toolkit, which consists of five items. These are **five tools or core concepts** that you will need to chart a productive path on your journey into MRI, as depicted in figure 1.1.

Figure 1.1. The five core concepts: (1) MRI incorporates waves, (2) MRI is a water image, (3) MRI has 'Multiple Looks,' (4) MRI in terms of electricity and magnetism. Changing electric field → makes magnetic field, and changing magnetic field → makes electric field. (5) MRI relies on dipoles and precession.

---

[3] For research in domains of cognitive science, neuroscience and psychology, there are many important topics of investigation for which there are now academic departments that have invested in managing their own MRI scanners [3, 4]. Education that assists learners from clinical, translational, and basic science researchers who may come from different areas across the world may also receive benefit from reading this book. There are so many areas of study that find applicability in MRI that societies such as the International Society of Magnetic Resonance in Medicine (ISMRM) were formed to respond to the clinical and research needs of the many multidisciplinary users.

1.01 Scaffolding Box: overview of core concepts

This book is intentionally designed to be read by many different types of learners. Whether you are a resident, technologist, MRI scientist, and/or anyone else curious about MRI, there may be different paths to how you would like to receive this information. Some readers may be already integrated into clinical environments and prefer to begin with content that feels closer to the 'big picture' rather than establishing ideas through a theoretical framework. In that case, those readers could start their reading by beginning with a later chapter[4] and refer to the earlier chapters when they would like to engage with this content.

As mentioned in the 'how to use this book' section in the preface, readers will hopefully find that these concepts provide a missing link that will help them build knowledge. We highly recommend at least looking over the interactive components, particularly on waves, in this chapter, if that is not already familiar. The content in this chapter will introduce the core principles in a logical fashion to help all learners. This is useful whether they are looking backward in approach by reading later chapters and then coming back to earlier core ideas, or if they are developing a forward approach by starting with the fundamentals in this chapter. Please feel free to continue with items in this chapter or start in chapter 2 or chapter 3, whichever works best for you.

MRI Made Easy: (...well Almost) by Hans Schild [5] is radiology resident- and/or technologist-friendly and uses a cartoon illustrative style that is very 'adjunctive' to chapter 3. We also recommend everybody beginning with MRI to read it in conjunction with this text.

However, please consider the following ideas (figure 1.1), whether you are reading on throughout this chapter or if you are starting at a different point in this book:

(1) MRI relies on electromagnetic (EM) waves. Different waves have frequency and phase offsets, which are important to pulse sequences and artifacts in images later in this book.
(2) MRI is water-centric in terms of clinical imaging. Abnormal alterations in water content can be signs of disease.
(3) 'Multiple Looks' can provide multiple contrasts that guide more confirmatory evidence for a diagnosis. At the same time, multiple looks can lead to greater complexity and require greater experience and understanding when applying MRI.
(4) MRI relies on imaging that results from magnetic dipoles and precession of these dipoles. The precessional frequency is known as the Larmor frequency.
(5) $\Delta E \rightarrow M$, and $\Delta M \rightarrow E$[5] is thematic in the understanding of MRI, especially in terms of excitation and reception of signal.

These core concepts are discussed in the following subsections of this book.

---

[4] In the preface of this book, there is a chart and suggested table of where different types of learners and professionals may chose to use this book.
[5] The Greek symbol Δ is shorthand for a 'change' in value. This symbol is widely used in chemistry, mathematics, and physics, and will be used throughout this book.

Figure 1.2. Chapter 1 will describe these five core principles useful for understanding MRI.

We have created the following symbols that are used to remind you to reflect on the five scaffolding ideas shown in figure 1.2. These symbols will serve as a potential guide to remind you of these five 'core' concepts of this chapter.

This book is attempting something that we wish we could do for you in person—to provide a strategy of learning that meets you where you are as a learner. In this chapter, we will introduce the core concepts. Each concept will be indicated by a corresponding symbol; in later chapters, these symbols will indicate when material is related to one of the core concepts. When you see one of the symbols, you may wish to refer 'back' to the list of core concepts and make sure you understand how the content is related to these core ideas. Boxed items (□) are meant more to provide a small aside, point out fun or interesting facts, provide historical context, and/or suggest further learning scaffolds which in turn may help preview some ideas needed for understanding complex topics. As 21st-century learners, you have many ways to get to information. Thus, we also encourage you to cruise the web, follow up on references, search out instructional videos, and so forth. Use this book as a guide to form a framework for understanding MRI, whether you are a resident, medical student, technologist, or interested in research.

On a final note, we will apply a deductive approach to train your observational skills, and to develop your critical thinking. We will call these 'Deductive Learning Scenes,' or DLS, which will be marked by the 'detective' magnifying glass.

For these sketches, please write down what you observe in the sketches/scenes. Then, look at the suggested answers that we provided. The goal is to be able to help you through the sketch to help you retain the evidence in the image. Observe, while paying particular attention to the five core concepts that can enhance your framework of these concepts.

This exercise matches more closely with the medical model, in which you are presented with a complex model, and you must deconstruct and analyze it in an organized fashion.

In this section, we continue to discuss five core concepts of understanding for MRI. Please note that in this book, learners are not expected to derive mathematical or physics-based answers or even memorize formulas, but rather are encouraged to focus on understanding the conceptual underpinnings of MRI physics. It is, however, helpful if you have had previous prerequisites in chemistry and physics that are required for a medical school or radiologic technologist degree. We believe and have observed that the more you can organize the information that you are learning and connect with more concepts, the more you may gather some longer-term benefits. You may first try to simply memorize facts at the start (one could equate this with first starting to learn vocabulary when one tries to acquire a new language, and it is very much part of the learning process), but as you progress, you will develop conceptual skills that may also help you retain more complex facts and information (in the learning language analogy, this could be understanding culture and specific nuances of a society). Similarly, in terms of medicine, the nuances and ideas can be differential to clinicians in the hospital realm. These conceptual skills at a minimum can be helpful to serve to organize the vast 'forest' of information needed for medicine. Maps and topics help us navigate the knowledge; we are also hoping that these frameworks serve to assist you to understand and retain the key information on the very elegant but complex topic of MRI.

In our program at OUHSC, the diagnostic radiology residents are allotted a relatively limited amount of time to study each modality and each anatomical domain (neuro, vascular, musculoskeletal body, etc), so the exposure to the MRI curricula is spread out over many years. It is especially helpful to develop problem-solving skills as you move from the first year in your training program through the second year. In the third year, it is beneficial to focus on flashcards and practice problems. Depending on your prior exposure to the material, during residency you may find that the information includes a fair bit of review of already-familiar concepts, even from high school or college math and science coursework. If your prior exposure is more limited, you may find that it is challenging to recall older materials, but once you learn some of the reasons why these concepts are important, you can seek to bolster and review your knowledge in that domain.

At this time, let us work through our first DLS, which is shown in figure 1.3. The following are observations from fellow residents and technologists.

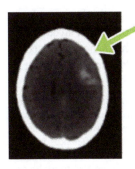

Looking initially at the CT scan...
1. We see the high density (bright values on the CT image) located within the left hemisphere of the brain (left is defined with reference to the patient orientation, which would be right from the point of view of looking at the page).
2. On CT, a high density (bright on image) would demonstrate a hyper-dense signal in the brain, indicative of a hemorrhage.

If we stopped there, we might send the patient to be treated with stroke clot busters (blood thinners).

Now let's turn our focus to the MRI scan….

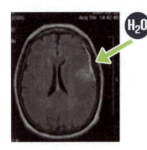

1. Notice the 'long' enhanced edge of the brain, which may resemble a tail. This is known in MRI as the '**dural tail enhancement** signal'. Enhancement of the dura/meninges is indicative of a meningioma, a tumor of meninges [6].
2. As MRI is a 'water' image, locations where **water shifts**[6], accumulates, and/or reacts differently can create strong contrast in the image. The treatment for tumors would lead to different outcomes for this patient than if the diagnosis was a hemorrhage from a stroke. Some of the treatment could lead to potentially dangerous and/or unnecessary interventions if made in accordance with an incorrect diagnosis.
3. The **MRI** illustrates that it has vital potential in patient care and in this case has provided some critical diagnostic evidence.

Dr Justin N. (Radiology Attending) reflects on the DLS in figure 1.2 above:
"I think it is important when looking at diagnostic images to realize that knowing what you are looking at is a skill and takes practice. It is a synthesis of knowing anatomy and pathology, but also knowing how the image is generated, so you know what the image is trying to tell you. It can be frustrating to face so much data head on... but keep looking! Each image you see teaches you something."

At this point, we have labeled items on the images below to guide you to a better understanding of this DLS[7], as shown in figure 1.3, and to learn more about anatomical structures in the brain.

The example above illustrates the power of MRI to image the brain and soft tissues in figure 1.4. **Soft tissues** connect, support, or encompass other structures in the body. **Soft tissue** refers to all of the tissue in the body that is not hardened by the processes of ossification and/or calcification. MRI excels at soft tissue contrast

---

[6] For this book, we discuss '**water shifts**' to describe the effects that move water from one part of the body to another. There are differences in water between intracellular space and the extracellular space that can vary due to the functioning of active sodium–potassium pumps, as well as the distribution of water during ongoing inflammatory or wound healing processes. Signal changes that are expressed due to differences in water content on the images can be indicative of the presence or absence of disease.

[7] This Deductive Learning Sketch (DLS) may be less obvious to researchers, as it is targeted to health care professionals. It illustrates that MRI provides unique presentations that more clearly aid diagnosis, and thus is an important clinical modality.

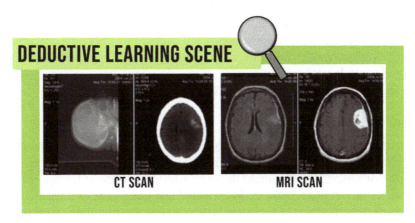

Figure 1.3. DLS comparison between CT scans and MRI scans.

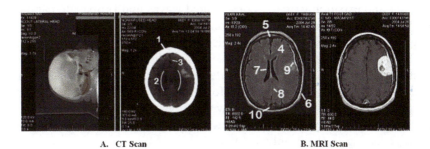

Figure 1.4. Labeled anatomy for the DLS sketch; (A) (1) skull, (2) centrum semiovale, (3) longitudinal cerebral fissure (B) (4) frontal lobe, (5) longitudinal cerebral fissure, (6) scalp, (7) lateral ventricle, (8) falx cerebri, (9) meningioma dural tail, (10) sagittal sinus. Adobe Stock Images: © VectorMine.

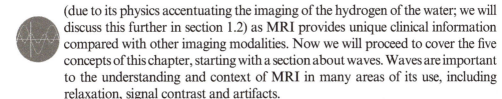 (due to its physics accentuating the imaging of the hydrogen of the water; we will discuss this further in section 1.2) as MRI provides unique clinical information compared with other imaging modalities. Now we will proceed to cover the five concepts of this chapter, starting with a section about waves. Waves are important to the understanding and context of MRI in many areas of its use, including relaxation, signal contrast and artifacts.

1.1 Waves

Learning objectives:
1. Understand two types of waves: transverse and longitudinal. Relate the transverse form of wave to E&M waves.
2. State several sources of E&M waves.
3. Understand frequency of a (transverse) wave.
4. Conceptualize energy relationship to frequency of wave.
5. Conceptualize that the decay of wave strength could be due to inherent natural causes.
6. Understand the phase of a (transverse) wave.

7. Understand the definitions of **bandwidth** and **spectrum**.
8. Look at a real-world application of spectrum in magnetic resonance imaging.
9. Conceptualize the addition of waves from several frequencies (sometimes called synthesis).
10. Describe what the Fourier transformation is and what it does.

1.1.1 Two types of waves

It is valuable to quickly understand two different forms of waves: some that move through a medium or material and some that do not need a medium. A wave transports energy from one location (its source) to another location without transporting matter. The two types of waves of focus are **longitudinal and transverse**, as illustrated in figure 1.5, which are classified by the vibration patterns and direction they travel in.

Longitudinal waves require a physical medium, such as a solid, liquid, or gas, for propagation. To picture longitudinal waves, imagine particles rhythmically compressing together and expanding apart. Longitudinal waves that propagate parallel to the particle vibration are important to ultrasound, but are less commonly relied on in MRI for imaging[8]. Note that MRI relies on the electromagnetic (EM) waves, which are of the transverse form of the wave. The transverse wave is discussed in the next paragraph.

In **transverse waves**, the particles vibrate perpendicular to the propagation of the wave. They can utilize a solid medium for travel or be self-propagating, meaning they require no medium and can travel within a vacuum. This is the case with EM waves. MRI utilizes only EM waves, so we will focus on EM waves for the rest of this book, referring to them only as 'waves'.

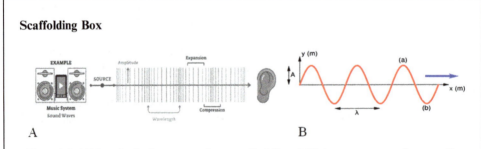

Figure 1.5. (A) Longitudinal waves are shown on the left, and (B) transverse waves shown on the right. Note that EM waves that are used in MRI are transverse waves. If you interested in reviewing more about these two types of waves, please look at the following reference [7].

---

[8] One exception for the use of longitudinal waves in MRI is in the domain of magnetic resonance elastography (MRE). MRE integrates MR imaging with a method that stimulates low-frequency vibrations in the patient. Such techniques create a visual map that can illustrate different material composition of body tissues. MRE can be used to detect stiffening that could evolve from fibrosis, wound healing, and inflammation in pathology, such as in chronic liver disease and/or in breast disease.

1.1.2 Electromagnetic (EM) waves

MRI uses EM waves to create the signal used in the image. Gaining a fluid understanding of the transverse wave can improve your comprehension of MRI in its reliance on waves. Electromagnetic radiation can be described as both an electrical and magnetic disturbance traveling through space at the speed of light. Examples of EM waves include light (from the Sun), x-rays, and microwaves (such as those in your kitchen microwave oven). For those who are curious, consider some basic material as a brief tour/review of the concepts [8].

1.1.3 Mathematical properties of transverse waves

At this point, let us look at the wave and review some of a wave's mathematical properties. First, the height of the wave goes up and down, creating a peak-to-trough effect. The amplitude of the wave is from the x-axis to the peak, as shown in figure 1.6 (also equivalent to the distance from the x-axis to the trough). A second feature is the period, defined as the peak-to-peak spacing of each repetition (trough-to-trough can be used also, as it is identical). The 'period' of the wave is also referred to as the wavelength. When we decrease the space between each period, we are increasing the rate of fluctuations of the wave. The decrease in period also increases the frequency of the wave[9]. Note this form of wave is for the **spatial wave** form [7].

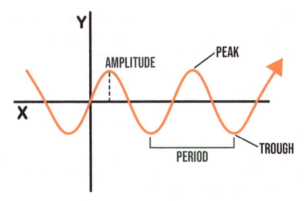

Figure 1.6. Illustration of the components of a transverse wave. **Amplitude** = distance between the peak and the y-axis of a wave (in y-direction): **Period** = distance between two peaks or troughs in a wave (in x-direction)[10].

---

[9] For further insight on waves, you can also watch videos such as those at Introduction to waves (video) Khan Academy [7].
[10] Waves are patterns that can occur in space and/or time. The reader may encounter either 'time' or 'space', such as distance on the x-axis. Amplitude is usually on the y-axis and may physically represent different features, such as distance, energy, etc [13].

As you see in figure 1.6, the wave contains a repetition of cycles. As the wave *propagates* from left to right, the distance from the *y*-axis *oscillates* up and down, from peak to trough. This displacement is called the amplitude of the wave.

A second feature (other than **amplitude**) is the spacing between the peaks. Note the spacing in a wave between peaks is the same as the space between troughs. The **wavelength**, or **period**, refers to one cycle and is measured from one peak to the next. The period of a wave is measured in units of time, such as seconds that it takes to complete a cycle in seconds/cycle. In mathematical terms, the frequency is the inverse of the period; it is measured as the number of complete cycles of a wave in a given amount of time. Units for *frequency* could be reported in hertz, which is equal to the quantities in terms of cycles/*second*. The frequency has an inverse relationship to the period of a wave, such as described by the following equation[11]:

$$\text{Frequency} = 1/\text{period} \qquad (1.1)$$

As a final note, there are two different types of frequency: **spatial** and **temporal**. Depending on what is represented on the x-axis, which can be **space** in some cases and/or **time**, then 'frequency' will have units of inverse distance (cycles/cm) and/or inverse time (cycles/sec). The reader will have to carefully consider the context (space and/or time) in which a wave is represented. In space (spatial frequency is known as *k*-space in MRI) and in time, like we describe precession and how it works, we refer to a temporal frequency (i.e., time-based frequency).

1.1.4 Understanding energy of waves in relation to frequency of waves

Inspect the image below and write down what you observe in the following deductive learning sketch. Remember that a wave transports energy from one location (its source) to another location as sketched in the DLS example in figure 1.7.

Note in the above figure that the two waves have different frequencies. Energy within a wave is determined by both frequency and amplitude. Keeping amplitude constant, we can focus on frequency in that figure. The amplitudes in the above DLS are the same (height of the oscillations). However, the number of periods, or oscillations, differs. Examining the top of figure 1.7, the oscillations are more numerous, indicating that the top rope has a higher frequency and carries more **energy** when compared with the lower rope.

For a conceptual understanding of this relationship, think of the waves in a wiggling 'jump' rope. To generate a higher-frequency wave in a rope, you must move the rope up and down more quickly. This takes greater energy; thus, conceptually, a higher-frequency wave has more energy than does a lower-frequency wave.

---

[11] Sometimes you might see frequency expressed as **angular frequency** (in radians per second). In this book, we will use **frequency** to be measured in units of Hz (in cycles per second, which is called hertz or Hz). Note that the difference between **angular frequency** and **frequency** is a factor of 2π. This is because 2π rad/s corresponds to 1 Hz if you are ever asked to convert units between the two systems of expressing frequency.

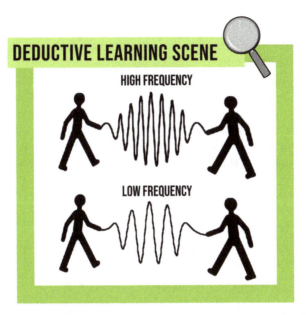

Figure 1.7. DLS highlighting the relationship between frequency and energy. Note that the amplitudes of the two waves are the same in this example.

1.1.4.1 Fun-fact bubble: a fun experiment thinking about energy and wavelength with a jump rope

If you have a rope available, you can do an experiment on what (energy) is required to generate higher frequencies. Can you describe why it would seem to take more energy to generate faster frequency?

(a) Take a rope and tie it to a tree or a post. (Use a 5-ft-length rope for convenience, but one that is at least long enough for you to generate waves.)

(b) Oscillate the rope by shaking it up and down. Now, keep the amplitude of the wave about the same by creating a higher frequency wave. Moving closer to or further from the tree can change the amplitude of the wave[12].

If this seems trivial to you, then show your kids and/or your nieces/nephews as a fun assignment to have them get a feeling of waves. If you haven't already, once you begin to start thinking more about waves, it will help you become more insightful when it comes to MRI. This is especially the case for the more complex topics, such as signal to noise and time tradeoffs (chapter 6) and imaging artifacts (chapter 7), that arise in day-to-day clinical practice.

---

[12] This rope image has been obtained by the author from the Pixabay website where it was made available under the Pixabay License. It is included within this book on that basis.

1.1.5 Decay/damping of waves

Many things in nature have a decay and/or attenuation of that wave, as seen in figure 1.8, which is created through a dampening effect. The physics definition of the term **dampening** means reduces the amplitude of a signal. Consider a bouncing ball that bounces less and less, and eventually stops bouncing over time. An exponential decay envelope can be used to model more natural decay elements of a wave as depticted in figure 1.8[13]. Dampening can arise when you are sending energy that is not traveling in a vacuum.

Self-propagating waves can continue 'forever' in the absence of obstructions. Transverse waves can travel for long distances. In fact, some waves can seem to go on forever; think of starlight, which travels through the vacuum of space, but still reaches us, as portrayed in figure 1.9. Other waves (unlike starlight) have greater attenuation as they travel through space, much like the pond waves that emanate off the splashing tail of a beaver, as shown in figure 1.10[14].

Returning back to more about the attenuated signal. Mathematically, the exponential function can be written as:

$$f(t) = e^{-\alpha t} \text{ or alternatively, } f(t) = e^{-t/T} \tag{1.2}$$

where e is the natural exponent, a is a decay rate constant, and T is the time constant of decay. $f(t)$ is the change of the function's value as time (t) changes. Note that $a = 1/T$, so that the decay rate constant (a) is the inverse of the time constant of decay (T).

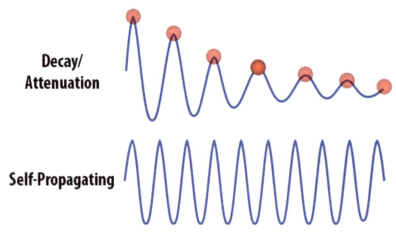

Figure 1.8. Dampening of a sinusoidal wave, note the exponential decay. Balls can exhibit an attenuating motion as they diminish each bounce over time when they are permitted to freely bounce.

---

[13] The enthusiastic reader may like to preview an upcoming topic known as Free Induction Decay that will be further described in chapter 4, by watching the following video from Dr Paul Callahan on Acquiring a Free Induction Decay (FID) [11].

[14] We have introduced the concept of the transverse wave with both electricity and magnetism (E&M) physics, as well as with mechanical waves. While the underlying physics appear differently due to how they are formed and the fundamental forces that underlie them, we still encourage the reader to try to think about these concepts in both the E&M world and the mechanical world. For further exploration, consider going to the hyper physics website, which is a great resource for refreshing physics fundamentals, http://hyperphysics.phy-astr.gsu.edu/hbase/Sound/wavplt.html.

Figure 1.9. Ever thought about starlight? We can even observe the light from stars in Alpha Centauri (closest known stars to the Earth) that has travelled over ~4.3 light-years from us. These waves travel through space and don't have inherent damping, at least not as significantly as we would see in other types of waves here on Earth that occur in Nature. Left image, credit: NASA/Penn State University; right image, this 'File:Alpha, Beta and Proxima Centauri (1).jpg' image has been obtained by the author from the Wikimedia website where it was made available User:Skatebiker under a CC BY-SA 3.0 licence. It is included within this article on that basis. It is attributed to User:Skatebiker.

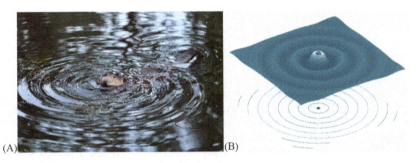

Figure 1.10. (A) A dampening wave generated by a beaver as it swims. This File:Beaver at Meadowbrook Pond, Seattle 18.jpg image has been obtained by the author from the Wikimedia website where it was made available by User:Jmabel under a CC BY 4.0 licence. It is included within this article on that basis. It is attributed to User: Jmabel. (B) A simulated 2D wave, further evidencing the motion and decay in multiple dimensions.

See figure 1.8 for the decay curve. Note that the time constant will be something of particular interest when we talk about T1 and T2 relaxation in chapter 3.

Now that you have reached this point in the waves concept, you have done an excellent job. The next step is to gain some interactive and intuitive practice. Please go to the PHET website [12]. Many learners have really enjoyed this simulation, particularly as it provides a visual intuitive sense of a physical wave, as seen in figure 1.11. It is worth trying the app at least one time for the fun of it to place yourself in the mood for understanding waves.

Instructions for this website at the time of publication of this book:
(1) Open the app on the PHET website [12] and select **'wave intro'**.
(2) Click **'water'**, turn on **'Side View'**.
(3) Select **graph** checkbox in the upper right.
(4) Hit the **'green round button'** on faucet to turn on the water and start it dropping and creating waves.
(5) Slide the **amplitude** and **frequency** back and forth to see what happens.
(6) Observe properties of wave and decay.

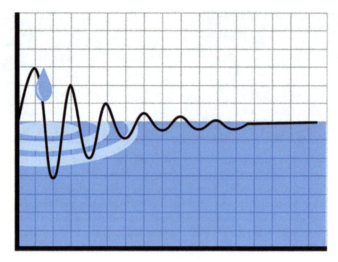

Figure 1.11. PHET simulation for altering properties of a wave, available at https://phet.colorado.edu/sims/html/waves-intro/latest/waves-intro_en.html [16].

Natalie (premedical biomedical engineering student) reflects on attenuation and waves:
"I have always thought about attenuation as you mentioned as illustrated with the ball, or also like a ripple in the water that slowly decays out. Like when you are fishing and make your first cast that starts a ripple that slowly fades out. I personally played basketball and softball in high school, so I also imagine figure 1.8 as when you shoot the ball and then it first hits the ground and fades out. I never gave much thought about self-propagating waves. I am curious to see how both wave properties will impact signals in MRI."

Dr. Wu added:
"That sounds like you are understanding decaying waves. The reason for the discussion of decaying waves is a preview for understanding 'relaxation' in MRI. Properties such as decay and/or attenuation in waves are ideas that are particularly important when we discuss the key concepts of T1, T2, and T2* relaxation. We will go into relaxation in greater detail in chapter 3."

1.1.6 'Phase versus frequency'

Let us look at the following deductive learning sketch, focusing on differences between phase and frequency of the blue, green, and red waves. Note: in this example, assume all waves have the same amplitude (i.e., their maximums and minimum

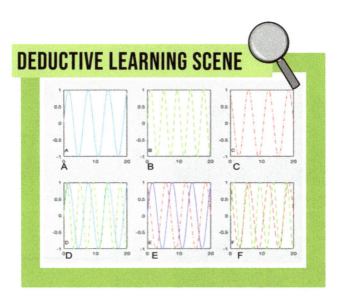

Figure 1.12. DLS comparing properties of waves. The reader should try to develop 'interpretations' of the differences between sets A, B and C in the first row. The second row exhibits comparisons between curves shown in the first row, which the reader would be asked to interpret. 'D' shows comparison between A and C. 'E' shows comparison between A and C. 'F' shows comparison between B and C.

values are the same in scale). The DLS sketch is shown in figure 1.12. For the exercise here, look at the second row (D, E and F) and try to guess what the differences are between the two curves (these are drawn from A, B, C) that are overlapped in pairs. In the first row, singular waves are shown for clarity, and are used in the second row of the DLS.

Answers or things to consider after performing the exercise from DLS of comparison of phase and frequency from the graph:

1.1.6.1
In figure 1.12D, the figures show **frequency difference** between the blue line in figure 1.12A and dotted green lines in figure 1.12B. Note that the peak-to-peak space is different between the green dotted line and the blue solid line.

1.1.6.2
In figure 1.12E, the frequency is the same between the blue figure 1.12A and the red figure 1.12C, but the **phase is different**. You can note that the peak-to-peak spacing in the wave is the same, but the starting positions are slightly offset.

1.1.6.3
In figure 1.12F, just provided for completeness in terms of comparisons (between case B and C), **both the phase and the frequency are different** in these cases.

1.1.6.4 Fun-fact bubble: thinking about cars and waves

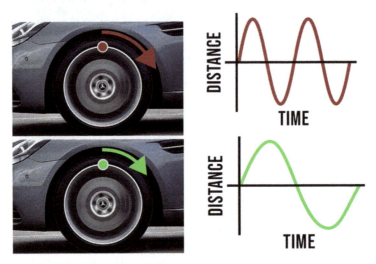

Figure 1.13. Comparing period and frequency using the cyclic motion of a wheel on a car. This image has been obtained by the author from the Pixabay website where it was made available under the Pixabay License. It is included within this article on that basis. The reader may be able to use the animation to better gain an intuition of both frequency and phase by considering aspects of rate of rotation and the start point of the oscillations. Animation available at http://www.youtube.com/watch?v=G5_zul5wrTY.

Say your friend has a car, as portrayed in figure 1.13. Looking at one of the tires, imagine a point at the top. As you drive, the tire begins to 'cycle' and the point makes a full rotation with the tire. This is the 'period.' The 'frequency' of the period is directly correlated with the speed of the car. Note that the 'red' car has a faster frequency of rotation than the 'green' car because the red car is traveling at a faster speed. Additional resources can be found at [14].

Practice thinking questions
1. Can you describe frequency?
2. Can you describe two forms of waves: transverse and longitudinal?
3. Can you describe what is meant by phase?
4. Can you conceptualize energy relationship to frequency of wave?
5. Can you conceptualize the decay of wave strength as an exponential curve?
6. Can you tell the difference between amplitude, frequency, and oscillation?

1.1.7 Understanding frequency spectrum and frequency bandwidth

You may remember from your classes the concept of the visible light spectrum, which consists of a small range (380–740 nm) extracted from the entire spectrum of electromagnetic wavelengths. A similar concept with audio waves allows us to tune

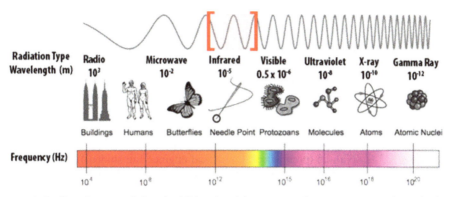

Figure 1.14. Radio, microwave, infrared, visible, ultraviolet, x-ray, and gamma rays are shown in the EM spectrum. Note that buildings, humans, a butterfly, a needle, a protozoa and molecule, and the nucleus of an atom are shown for reference as a scalar sense of size that can be compared with a 'wavelength' size. The reference exhibits a sense of spatial scale for the wavelength in this picture [15]. Note that a **bandwidth** (BW) is a range of frequencies of interest within a spectrum. For example, in figure 1.14, we marked the spectrum in the infrared range between 10^{-5} to 0.5×10^{-6}, but this range can be anywhere in a spectrum and of any size. Another example that is relevant for use in MRI is a bandwidth of 120 KHz. For example, a frequency range of 63.74 to 64.86 MHz could be a bandwidth of frequencies that is used in pulse sequences for a 1.5T Scanner (i.e., has center frequency ~63.8 MHz)[15]. This 'File:EM Spectrum Properties edit.svg' image has been obtained by the author from the Wikimedia website where it was made available User:Inductiveload under a CC BY-SA 3.0 licence. It is included within this article on that basis. It is attributed to User:Inductiveload, NASA.

in to specific signals from specific radio stations. This analogy will be used in the book to remind us that MRI scanners are not observing at a single frequency, but at many frequencies across a bandwidth. For this book, the spectrum of a signal is a set of (continuous) frequencies contained in the signal, as rendered in figure 1.14. The bandwidth refers to a range between a lower bound and upper bound of frequencies in the EM spectrum. See figure 1.14, which demonstrates both spectrum and bandwidth. Recall that waves carry energy. Waves can also carry data by converting energy to said data. These data can be auditory—as in music from a radio, or visual—as in an MRI image.

Sound waves from a radio station are altered to travel further when transmitted. When received by an antenna, the radio converts the wave back to audio. Similarly, protons give off energy and the frequency of the energy waves is received by MRI coils[16]. A computer can then convert the waves to an image using a mathematical transform known as the Fourier transform, which is depicted in figure 1.15.

---

[15] The Larmor frequency is 63.8 MHz for a 1.5T magnet. A bandwidth of 20 kHz could represent a range of frequencies that expands around that central frequency such as the Larmor Frequency [63.79 MHz, 63.81 MHz]. The bandwidth is the magnitude of that range (i.e., 20 kHz).

[16] A key concept here is that the MRI acquires the signal in the frequency space (spectrum) and that by algorithms these sets of frequencies and their corresponding amplitudes are used to form the signal as we describe in section 1.1.13.

Amy B (psychology graduate student) reflected about spectrum and bandwidth:
"When I think of wavelength, I think of vision, and that seems to happen in the specific set of continuous frequencies. When I think of bandwidth, I think of it in regards to technology."

Natalie (premedical biomedical engineering student) also said:
"I think I can visualize what you are saying, I think I'll read on and figure how bandwidth and spectrum are used in MRI, but that gave me a good insight into the topics."

Figure 1.15. Sound is encoded using a wave with varying frequency and converted back to sound by a radio. This image has been obtained by the author from the Pixabay website where it was made available under the Pixabay License. It is included within this article on that basis. Similarly, energy waves with varying frequency are released from protons and interpreted to form a MRI image. *K*-space image on the left depicts the varying frequencies that make up the portrait of Vincent van Gogh on the right[17]. This 'File:Vincent van Gogh - Self-Portrait - Google Art Project.jpg' image has been obtained by the author from the Wikimedia website, where it is stated to have been released into the public domain. It is included within this article on that basis.

---

[17] When you hear about MRI, you will sooner or later come across the term *k*-space. We introduce the concept here, but it is additionally covered or used in chapters 4–7. Do not be too bothered by the term now. I have found that many people from all sorts of disciplines—technologists, chemists, psychology, medical students—are able to understand some of the core components of k-space after some time spent with the concepts in this book.

1.1.7.1 Fun-fact box: radio stations and understanding bandwidth

In the United States, frequency-modulated broadcasting stations operate between 87.8 MHz to 108 MHz, for a total of 20.2 MHz. This range of frequencies is an example of bandwidth. We will further review bandwidth in chapter 5

1.1.7.2 Applied bubble

Real-world application

Each element interacts uniquely with frequencies based on nucleic properties such as charge and spin.

NMR spectroscopy utilizes these characteristics to identify molecules by their chemical structures. Clinically, NMR spectroscopy can be used to identify metabolites indicative of a diagnosis, such as abnormal cell growth due to a brain tumor, as shown in figure 1.16.

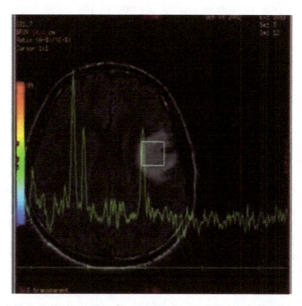

Figure 1.16. A single voxel is scanned at a specific frequency. This is an MRI of a brain tumor. The spikes in spectral data identify various functional groups that are pieced together like a puzzle to identify present metabolites. The area under the curve reflects the amount of chemical makeup in a part of the brain. This information can help stage and/or monitor the cancer.

Carlie (premedical student, with chemistry and psychology emphasis, and dancer) reflects on NMR in her former chemistry class:
"I remember using NMR spectroscopy in my organic chemistry class to verify if certain products were present after reactions. After acquiring our spectrum, we would measure the area under the peaks, which would give us information about the structure and constituents of the compound. While NMR is a staple in the chemistry lab, it is just as important in the clinic because its multidimensional application led to the development of MRI and the ability to observe tissue matter like the brain or heart in a non-invasive way."

1.1.8 Adding different waves together (a form of synthesizing waves)

The reason we are introducing this topic is that MRI utilizes an acquisition process that collects multiple simultaneous waves at different frequencies that are received by the MRI hardware. MRI encodes each point in space in both frequency and phase, and then sends the sum of all those waves to a receiver coil (more on this topic will be covered in chapters 2 and 3). The term 'synthesizing' waves that we will use in this chapter comes from the audio world and is used to describe the concept of adding a bunch of waves with different frequencies and phases to create a new wave (figure 1.16). We have included interactive exercises that one can do as a DLS in figure 1.18.

This may provide greater intuition regarding the synthesis of waves. Otherwise, please read on, as this section is setting up one of the most important concepts in MRI, the Fourier transform, which is discussed in the next subsection, 1.1.12.

Nick (third-year medical student) said:
"The synthesis of the waves seems to be summing up the separate waves' frequencies and amplitudes to a resultant summarized wave, which is fortunately done automatically by the MRI machines. The Fourier Transform seems to be an equation that does the same thing as the MRI, but adds up each individual wave by hand."

To help understand the process of synthesis (or acquisition) in MRI, we will look at two examples. Figure 1.17 above provides an analogy for wave synthesis. Swells in the ocean are naturally occurring due to varied wind patterns and physical interferences. Opposing winds influence the ocean currents, making it very unpredictable and choppy because the currents interact both constructively and destructively. Figure 1.18 demonstrates the same phenomenon from a mathematical viewpoint. Every point within a wave has a value, or slope. Each simultaneous, individual point within a wave can be added to create a new wave.

Next, let us move on to an exercise that provides you with an interactive experience that students have found useful, that represents two different waveforms as a combination of subcomponent waves that each have different amplitudes, as shown in figure 1.18. Note that the final result (green wave on the right of I and II) can look dramatically different.

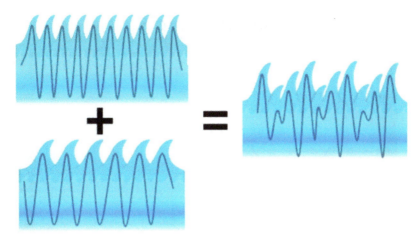

Figure 1.17. Multiple waves that are simultaneous will be perceived as a single wave. The 'addition' of waves has both constructive and destructive interference. This technique can be used intentionally to create unique wave patterns. Imagine that you can keep on adding more and more waves together to make a myriad of different shapes.

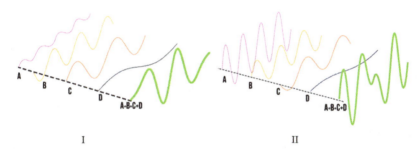

Figure 1.18. Cases I and II illustrate the same frequency and oscillations (i.e., four selected from a subset of frequencies). In A, we have altered the amplitude so that the sum of all waves with their amplitude results in a different synthesized wave. The amplitudes of each of four waves shown on the left axis (dotted line) are 0.35, −2.0, 2.3, and 1.0, respectively. The waves are then summed (additive synthesized) into the final form (green pattern shows over a 0 to 2pi range). Note that the frequencies of the waves are 4, 3, 2, and 1. In B, the amplitudes of each of four waves shown on the left axis (dotted line) are 2, 1.2, −1, and 0.3, respectively. The waves are then summed (additive synthesized) into the final form (green pattern shown over a 0 to 2pi range). Note that the frequencies of the waves are 4, 3, 2 and 1, respectively[18].

We have included interactive exercises that one can do as a DLS in figure 1.19. This may provide greater intuition regarding the synthesis of waves. Otherwise, please read on as this section is setting up one of the most important concepts in MRI, the Fourier transform, which is discussed in the next subsection 1.1.12.

---

[18] Note, in case I on the left wave, 'A' is the same as in case II, except that they are scaled with different heights. If you end up adding all the waves on the left to produce the green result in case I and compare it with the green result in case II, observe how you can generate almost an endless number of shapes if you just scale each subwave (purple, yellow, orange, blue) differently and add (synthesize) them together.

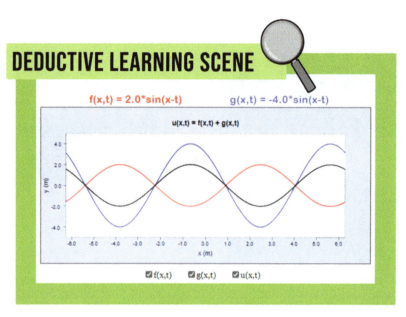

Figure 1.19. Synthesizing two waves together in the interactive online program at ComPADRE Digital Library website, https://www.compadre.org/osp/EJSS/4030/138.htm, similar to what is shown in this figure 1.17 [50].

Amy B. (psychology graduate student) reflects after looking at figure 1.18 and scanning ahead to figure 1.21:

"If the MRI is created, the 'synthesized' wave forms, and we know what frequencies were used to make those waves. It seems like the inverse Fourier transform is extracting out the individual amplitudes from those frequencies that synthesized the MRI signal. This was like we looked at A + B + C + D together in figure 1.18. The final image we get from the MRI is the ABCD synthesized frequencies that the inverse Fourier transform gives us based on the individual A + B + C + D amplitudes for the given frequencies. I think you are next going to show us the PHET wave game in figure 1.21, which I think is coming up in a following section of this chapter, and with figure 1.18 and the upcoming figure 1.21 it helped me solidify these concepts together about encoding waves."

Dr. Wu responded to Amy B. and said:

"Amy, I think you are getting why we put that figure 1.17 and connecting that with the later figures 1.18 and 1.19. It's helpful to relate to how we can encode images from the MRI machine. These MRI signals are spectral in nature and we then end up applying the inverse Fourier transform to decode those images into an image. I don't think these concepts come easy at first, but if you are able to walk through figure 1.21, which is coming up shortly, I think you'll be able to feel more comfortable with MRI concepts that are coming up in the later chapters."

Try to change the frequency between the two, and the phase and/or a combination of both. After you compare the waves, look at the addition of $f(x,t)$ and $g(x,t)$ by clicking on the $u(x,t)$ box. For example, enter the following functions for 'g,' function 'f' will remain constant.

$f(x,t) = 2.0*\sin(x)$
$g(x,t) = 1.0*\sin(x)$ ← different amplitude compared with $f(x,t)$
$g(x,t) = 2.0*\sin(2*x)$ ← different frequency compared with $f(x,t)$
$g(x,t) = 2.0*\sin(x-2)$ ← different phase compared with $f(x,t)$
$g(x,t) = 2.0*\sin(2*x-2)$ ← different frequency and phase compared with $f(x,t)$.

Amy B. (psychology graduate student) reflects:
"I get it, I'm adding waves together and you see the merged version of them, which is a synthesized wave. I felt the compadre site was also helpful in visualizing the synthesis. I'll keep reading more and see how we end up using this in MRI."

1.1.9 Fourier transform

In MRI, the Fourier transform decomposes any function of space into constituent spatial frequencies. Inverse Fourier transform can be also used to retrieve spectral information from an image. Because an MRI is constructed by waves that are combined together during the acquisition process, the resulting synthesized waves that are produced can be Fourier transformed into an image, as depicted in figure 1.20.

Gail H. (neuroscience postdoctoral researcher) reflects on the Fourier transform:
"Complex signals can be reconstructed by adding together simpler wave signals. The opposite is also true; you can break down a complex signal into the component waves that create it. The Fourier transform breaks down a signal into its component information, and the inverse Fourier transform reconstructs a signal using its components. Since we are talking about waves, the information inside the signal involves frequency, phase, and amplitude. When you start talking about k-space, that is getting into storing signal information about phase and frequency. You have to sample the information from k-space to reconstruct the image using inverse Fourier transform."

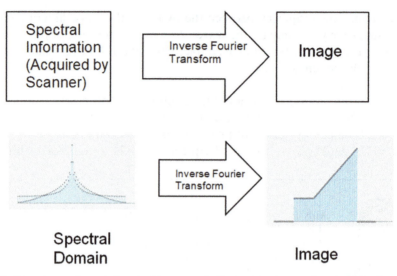

Figure 1.20. There is a mathematical method known as the 'inverse Fourier transform,' which will enable us to retrieve the values. The key here is to think in two spaces: the spatial map (y is amplitude and x is position), and the spectral space (y is the amplitude in frequency space, and x is the frequency values). While this figure may initially appear to be complex. the idea of encoding multiple signal values in a spectrum is one that will appear repeatedly throughout this book in several places (for example, sections 2.4 and 4.2),

---

1.1.9.1 Fun-fact bubble on mechanical Fourier transform

If you are curious and want to see the mechanical Fourier transform machine made from gears, please see the linked video. This mechanical Fourier transform machine can provide an intuitive feeling for the algorithm. The following 'job' of the Fourier Transform is to take 'synthesized' frequency signals and turn them into spatial coordinates [16].

A key concept here is that the MRI acquires the signal in the frequency space, which is a spectral wave, and that by algorithms known as the inverse Fourier transform [17] takes this signal, and this set of frequencies and their corresponding amplitudes are used to create the image. Another resource for understanding the Fourier transform is provided in the article by Gallagher *et al* in the *American Journal of Roentgenology* in 2008, for those who would like to see more details written in article form [18]. The goal is to provide the 'analysis or desynthesis'—in the latter, we're referring to the extraction of the components, such as the amplitude and the frequency and phase from the mixed signal that is retrieved from the scanner during signal acquisition.

1.1.9.2 Fun-fact bubble: Complexity of the wave domain: bats and dolphins
It may initially be a little hard to think in terms of frequencies. After all, we are not dolphins and/or bats! To help us recognize when we are seeing something in the spectral space, consider looking for one of these symbols[19].

While, it may almost seem impossible to grasp concepts like the Fourier Transform, please consider looking at videos. Consider the concepts of adding different waves together with different amplitudes and realize there are mathematical waves to extract these amplitudes. This mathematical method is known as the Fourier transform [18][20].

To try to have intuition for the Fourier transform without having to apply mathematical formulas, the reader should try to work at the DLS in figure 1.21. We believe that you will be on your way to some understanding of its 'magical' ability to extract images from waves of signals.

Please add the following after the figure:

Instructions for DLS in figure 1.21 above.
1. Go to PHET Fourier: Making Waves website [19].
2. Click Wave Game, and select a level (three would be a reasonable level to try.)
3. Move the 'amplitudes' slider bars up and down and observe the figure on the lower row, which shows the synthesized waves. The amplitudes will be your 'image.' The slider bars appear as three circular dots in a gray rounded rectangle.
4. Note that there is only one pattern of amplitudes (that you picked in the top figure) **that match** the final complex wave form. The middle layer represents waves that have the specified frequency with the corresponding amplitudes you specified. You should try to match the pattern.
5. Check your answer and/or show the answer, but it's worth a try to move the sliders up and down.
6. We do this in our lectures and it is well received. Give it an 'attempt' to match the waves. It may provide you with a sense of how you can encode images with waves.

---

[19] Dolphin image rom Dolphin silhouettes Stock Vector, Adobe Stock, jan stopka/stock.adobe.com. Bat 'File: Bat shadow black.svg' image from the Wikimedia website, where it is stated to have been released into the public domain. It is included within this article on that basis.

[20] For simplification, we describe the inverse Fourier transform (IFT) as reconstructing the MRI image from the spectral domain. The Fourier transform is the forward version of the transform, which is basically the encoding that the scanner apparatus employs to generate the spectral signal that will be decoded later as the image by the IFT. To learn more, see the Gallagher *et al* AJR publication [17].

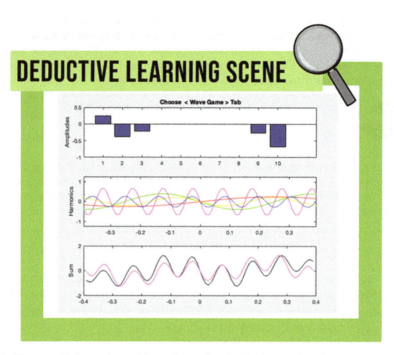

Figure 1.21. Please try this interactive tool for understanding the Fourier transform/inverse Fourier transform, available at https://phet.colorado.edu/sims/html/fourier-making-waves/latest/fourier-making-waves_en.html. It is highly recommended by several students, and users of this book found this simulation extremely useful [19].

1.1.9.3 Fun-fact box

The above is an advanced video for those who are both visual and enjoy a mathematical approach to understanding the Fourier transform [20].

While it could take some time to really understand the mechanics of the Fourier transform, please consider the following take-home points:

(1) You can add several sinusoidal waves with different frequencies and phase together.
(2) You can create any periodic function using a sum of many frequencies and phases of waves. This is the basis of the Fourier transform and what we will refer to as k-space in the later chapters.

Natalie N. (Pre-Med Biomedical Engineering Student) reflected:
"So that you are saying that you can encode information into waves and send it distances...then somehow turn that information into an image? I wonder how frequencies and phase could be used with that. That is so interesting, I look forward to learning more."

Dr. Wu replied:
"Yes, it may not appear obvious at first, as we aren't dolphins or bats, as we mentioned in 1.1.12.2. We must learn to feel more comfortable with waves and keep on working on it. We will see in chapter 2 how we can encode spatial information with hardware gradients, and in chapter 4 we cover frequency and phase encoding in pulse sequences, which center around those concepts you learned. As you get some practice, classroom time, experience, and understand the value of these concepts, it should impact clinical work and become more intuitive."

Final knowledge check:
- Understand frequency.
- Understand that there are two types of waves: transverse and longitudinal. The former is important.
- Understand phase.
- Conceptualize energy relationship to frequency of wave—can you relate this to the more advanced concepts?
- Conceptualize the decay of wave strength—did you try to relate this to the more advanced concepts?
- Conceptualize the exponential curve.
- Conceptualize a bandwidth of frequencies.
- Conceptualize a real-world spectrum, such as MR spectroscopy.
- Conceptualize the machinery of Fourier transform.
- Recognize that waves are not only important to MRI, but also to other modalities, such as ultrasound and even x-ray equipment.

1.1.10 Summary

Remember, waves are everywhere. Waves can have different frequencies. For example, your cell phone has a different frequency than your neighbor's cell phone, so that if you both need to make a phone call at the same time, the cell tower recognizes these different frequencies so that this is possible. This also means that waves can be combined. When combined, they can reproduce (or add up to) another wave. In the Fourier transform, waves with different amplitudes and frequencies from the spectral domain are added together (synthesized) to create a spatial image, like we see in MRI. Another real-life example: a note on a guitar makes a certain sound based on the waves. We can isolate (decompose) those waves so that we can see each component, and in turn we can then 'add up' (synthesize) those waves to recreate the note on the guitar. The opposite is true for the inverse Fourier transform; if we have a spatial image, we can 'decompose' the image into the separate waves (each with their own amplitude and frequency).

1.1.10.1 Yield bubble: a brief preview of 'k-space'

A majority of the MR image information is contained in the center of k-space, which we refer to as contributing to the 'contrast' of the image. Low-spatial-frequency data have the highest amplitude, which leads to largest changes in grayscale levels (altering the contrast). High-spatial-frequency data typically have lower amplitude in most images, as they focus on the details. Thus, the outside k-space contributes to the details of the image.

Here is a sample board question:
Which portion of k-space has lower typical gray scale values?
(a) Even lines.
(b) Outside portion.
(c) Inside portion.
(d) Neither; they are likely equal.
(e) Diagonal lines.

Answer: (b) Lower grayscale is lower amplitude. The details are more likely to have finer adjustment and lower amplitude than the major contrast.

To learn more, consider looking at this excellent Imaios resource [21].

Nick P. (third-year medical student) reflects on waves:
"It seems like waves are used everywhere in just normal life, including our phones, microwaves, even just seeing colors. It also seems to be integral to the workings of many of the imaging modalities used in radiology like MRI and ultrasound, and understanding the basics of how waves work and interact helps give a better understanding of what we are interpreting in the imaging."

Justin N. (attending radiologist) reflects on what Nick wrote:
"It's really interesting to hear about how medical students think about waves. I love how technology has forced people to have some concept of waves. Especially the microwave—it so clearly demonstrates how energy can pass through space and be detected or deposited. This basic understanding is very helpful when trying to teach concepts in MRI and ultrasound."

1.2 MRI as a water image

MRI uses magnetic fields and radio waves to measure how much water is in different tissues of the body; it then maps the location of the water and uses this information to generate an image. Because our bodies are composed of ~70% water [22], there is sufficient signal to produce detailed images (i.e., spatial and contrast resolution, as discussed in chapters 3 and 5).

 Carlie P. (premed and dancer) reflects about what she sees in the DLS in figure 1.22:
"In this picture, I see that the body is mostly made up of water, as well as features like the polar nature of water, and the spin of an atom. I'll read on to see why these elements are important."

Let's look at the DLS in figure 1.22. As you inspect it, write down what you see and describe in this image as depicted.

Below are items that you may notice and learn from the above **deductive learning sketch:**

- We are composed mostly of water molecules.
- There is H_2O: two hydrogen atoms + one oxygen atom. (Note these hydrogens give rise to the signal content in MRI/NMR [23]).
- Water has electrical polarity.

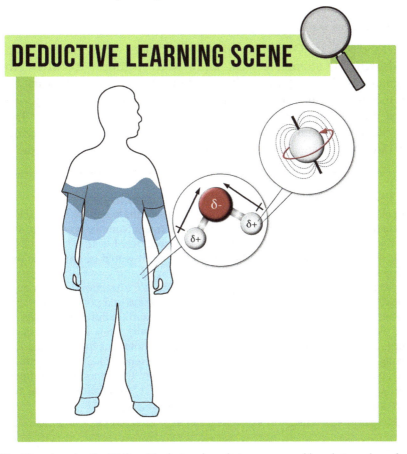

Figure 1.22. After observing the DLS and brainstorming what you see, consider what you learn from these observations. This is the interpretive phase of the DLS.

- We are concerned with the magnetic dipole associated with the hydrogen's nucleus, the proton, which is less commonly discussed. Protons are always spinning. The spin induces a magnetic field. Think of a magnetic dipole as just an idealized vector that functions as the pair of separated magnetic charges.
- There are waves, as illustrated in figure 1.24. (See waves within the figure, which was designed as a memory aid for your study.)
- Describe how 'water shifts' (such as edema) can be indicative of location of disease.

This leads us to the following **learning objectives:**
(1) Water is a large portion of the body, and the amount of water contributes to the amount of hydrogen that is needed to produce the MRI images.
(2) Protons spin, and have a **magnetic moment**[21]. This was predicted by the Stern–Gerlach Experiment, shown in the appendix.
(3) Electric dipole moments are shown, since you may have heard of these before, although we are most concerned with the magnetic moment.
(4) Describe edema and its relationship to water. Can edema be thought of as 'water shifts'?

1.2.1 MRI utilizes water

The water molecule (H_2O) is made up of two hydrogen atoms and one oxygen atom. The hydrogen (H) atoms are what make water pertinent to MRI. Protons, or hydrogen atoms, directly produce the signal measured by an MRI scan. Aside from one of the essential reasons why we can create clinical MRI images with these hydrogen protons, water is also miraculous due to the polar nature. This feature permits 'life' on our planet, as it prevents the oceans and seas from freezing solid all the way through (consider reading more in the 'fun-fact box' below in figure 1.24) [22].

1.2.1.1 Scaffolding fun-fact box: polarity of water and the oceans
The polarity of water has enabled life to arise from the ocean. Without the dipolar molecular nature of water, oceans and seas would freeze from the bottom up, and there would be no way for life beneath the surface of solid ice! Luckily, ice floats. In other words, if ice didn't float to the surface when it froze, it would be impossible for life to develop underwater!

Hydrogen bonding is a special type of dipole–dipole attraction between molecules. Ice floats to the top and gradually thickens freezing lakes and ponds (and other bodies of water, such as oceans and seas) from the top of the surface down. Hydrogen bonding (these electric dipoles interactions) is a key reason why ice is less dense than liquid water (i.e., if you put ice in a cup of water, you will note that the **ice floats**) and is an amazing substance that is essential to all life on our planet [24, 26]. See figure 1.23 for contemplative thinking that impacts life on earth and the importance of dipoles and water. *Life* on earth cannot survive without *water*. It is a precious natural resource. *Water* supports all human, plant, and animal *life* [24, 26].

---

[21] A 'moment' is a general physics expression product of distance and a physical quantity. In the case of 'magnetic moment' it is the amount of magnetism expressed over a distance.

Figure 1.23. The world's oceans are home to giant icebergs that float freely at the surface. This floating nature of water molecules is a unique property of the substance due to its polar nature [25]. Reproduced from Matthieu/stock.adobe.com.

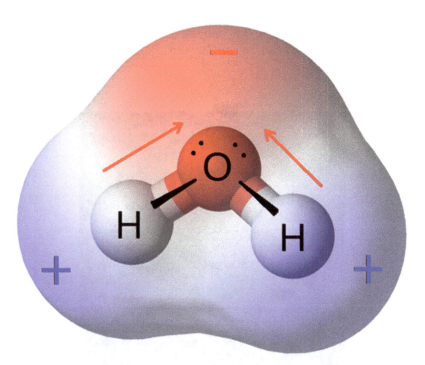

Figure 1.24. An interesting activity to examine a water molecule in more detail shown above (mostly review for those who have had a college chemistry class) [24]. Also, for those who remain curious, you may consider reviewing the chemistry of water blogs that are present on the internet [22].

1.2.2 Electric dipoles

Let us take a tiny detour and review a little about water. Water is made of both positively and negatively charged molecular portions, and actually has two electrical poles (one positive and one negative), as rendered in figure 1.24.

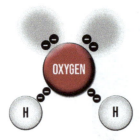

1.2.3 Water shifts

The presence of edema can indicate a response in wounds or damaged tissue that brings fluid to an affected area. **Swelling** is the result of the increased deposition of **fluid** and white blood cells in the **injured area**. An inflammatory response is associated with leaky vessels, which can be the cause of the swelling. 'Water shifts' to different anatomical locations can indicate tissue damage and can be used to detect disease using MRI, as can be seen in figure 1.25.

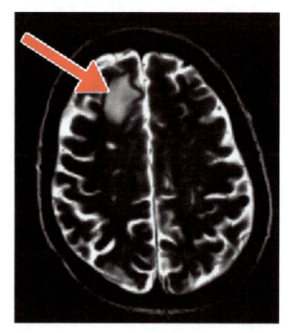

Figure 1.25. An image illustrating cerebral **edema**. **Cerebral edema** refers to a number of interconnected processes (including inflammation and trapped fluid) that can result in abnormal shifts of water in various compartments of the brain [27].

Summary

We are made of 70% water (a large fraction of our bodies is water). The electromagnetic characteristics of water are harnessed to generate the signal in MRI/NMR.

The spin of the proton plays a role in the MRI/NMR.
Waves play a role.
Notice the vectorial direction of the 'spin,' which is called a magnetic moment (produced by the magnetic dipole), in this case *g*.

Conceptualize protons as little tiny magnets that create magnetic moment.
Edema can be a response in the body for wound healing and inflammatory events. It can be thought of as 'water-shifts.'

So far, we have talked about 'mathematics' (of waves) and chemistry (of water). In the next section, we'll do some discussion in the medical decision-making field concerning 'multiple looks.'

Nick P. (third-year medical student) reflects on spins:
"So, the spin of the individual protons on each water molecule is what creates the image that we see on the MRI? I wonder how the MRI is able to convert the spin into a wave signal that is then used by the MRI machine to map the anatomy."

Dr. Wu answered:
"Nick, these are good starting thoughts on 'spin.' In section 2.6, we cover the receiver and, as the spin points to the receiver, the vectorial direction causes a changing magnetic field, which induces a current!"

1.3 Multiple looks (contemplating image weighting and the multiple appearances possible in MRI)

Learning objectives:
1. Provide a clinical description where multiple presentations aid a diagnosis.
2. Describe 'multiple looks' in terms of dimensions. Then, describe challenges.

You may have heard this saying before: If it looks like a duck, swims like a duck, and quacks like a duck, then it probably *is* a duck. This is what we will be referring to as multi-looks. It can be also thought of as multiple dimensions, multiple vantage points, **multiple weightings**, having a multifactorial nature, and/or several different lines of evidence. The way you see something. Different set of eyes on a problem.

Medical devices are designed to solve clinical problems. While there is a spectrum of possibilities for research, science, and exploration, we also hope that day-to-day these technologies can help us in the clinic. By having multi-looks, we have more ways of honing in on a correct diagnosis.

*Note: Multiple weightings may encompass a large majority of MRI, but in later chapters we will also learn the different ways we can create different 'looking' images (for example, perfusion, time of flight, diffusion tractography, etc). When we refer to **multiple looks**, we describe the advantages and possible disadvantages of having a number of vantage points (please note the discussion on the challenge of dimensionality below).

1.3.1 How do we achieve multiple looks/multiple weightings

In MRI we are able to achieve multiple 'looks' through the ability to select different pulse sequences to achieve different **multiple imaging weightings** (such as T1 or T2; see chapter 3 for more on relaxation and contrast weightings), FLAIR (chapter 4), and/or DWI, see figure 1.26. Again, this approach provides possibilities to have additional viewpoints—representations that will hopefully lead to better problem solving and more confidence in the findings.

For example, let us start with looking at a singular T2 image. Suppose you are attempting to make a diagnosis from this image in figure 1.27.

Now, examine when you have multiple weightings/looks at the same location.

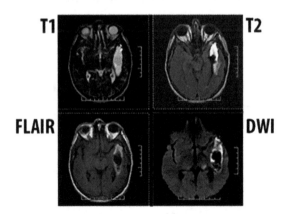

Figure 1.26. In this case, there are post-surgical changes, as well as the opportunity to identify locations of gliosis, hemorrhage, and the possibility of a recurrent tumor. Regardless, isn't it better to start with four looks instead of a single image? Multiple ways to get information provide a better likelihood of a correct diagnosis. MRI is versatile and complex, as it has several looks (T1, T2, etc).

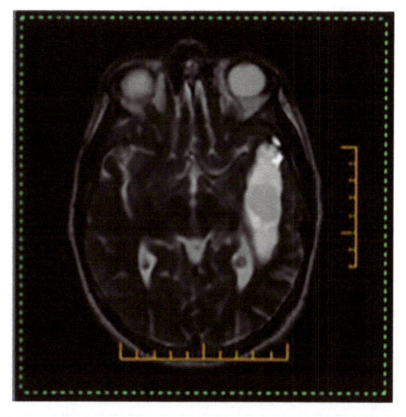

Figure 1.27. An example of a case where you can imagine you are trying to derive a diagnosis from just a single image that has just a single weighting. We will demonstrate the concept of 'multiple looks' in the following section.

1.3.2 Multiple looks and weightings as the analogy of different dimensions

MRI can be an excellent problem solving tool since it provides multiple looks. These multiple looks provide multiple dimensions of features (dimensionality is used here as another way to describe multiple contrasts or appearances), which permit the appearances that help the physician 'reader' to be more confident of their conclusions. However, note that multiple looks can add additional complexity. If you want to only see 'A,' then you must suppress B, C, and D as portrayed in figure 1.28.

Knowledge check:
(1) Understand the implication in MRI of having multiple looks.
(2) Can you come up with several ways of describing what 'multiple looks' means? How about examples in real life?
(3) Describe 'multiple looks' in terms of dimensions.

Possible Answer for (1): Having multiple images provides **confirmatory evidence** (more information from different vantage points). For example, with brain images, we end up being more certain that this is hemorrhage (as opposed to gliosis, etc) and its location in the image.

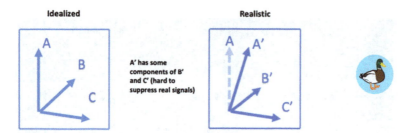

Figure 1.28. Challenge of realistic dimensionality within the context of multiple looks. Outwardly appearing images may include components even from the suppressed or reduced affects contained in other forms of weighting/contrast.

Possible Answer for (2): Multiple looks can be also thought of as multiple dimensions, multiple vantage points, **multiple contrasts**, multifactorial, and/or a number of different lines of evidence (restated from the text).

Possible Answer for (3): If you want to look at one dimension, then you have to suppress all other dimensions. For example, there are some consequences to interpreting DWI images with T2 shine-through, a topic that will be discussed in a future chapter.

We have now covered waves (mathematical), water (chemistry), and multiple looks (medical decision making). The next two subsections pertain mostly to chemistry and physics.

Multiple looks and multiple weightings is a tool that radiologists and research scientists use in their daily practice and provides a large advantage (with the slight curse of realistic dimensionality) in MRI. Note that some of the concepts occur in areas such as diffusion weighted imaging[22].

Nick P. (third-year medical student) and Justin N. (radiology attending) reflect on the concept of multiple "looks":

"We really like the explanation of the importance of multiple looks. As clinicians, this concept will be key to recognizing certain pathological processes like acute ischemic stroke and/or trauma in MRI, where standard images might not show any change at earlier stages."

"Every 'look' I get is a new piece of information. With a single perspective, I may not be able to provide specific data or diagnosis. An X-ray may get me to normal or abnormal, but I may not know the exact problem. But, the X-ray data can lead me to my next step, like CT or ultrasound."

---

[22] According to one of our residents, In a sense we are previewing the concept of T1 and T2 relaxation with the different appearances. However, when we refer to multiple looks, there are so many other variations that produce contrast. So, we lump all the multiple ways that an MRI contrast can appear in images into the term 'multiple-looks' for the reader.

1.4 Interlude and segue into concepts of electricity and magnetism

Up to this point in chapter 1, we have accomplished the review of three core concepts: 1) waves, 2) water, and 3) multiple-looks (see figure 1.29). The last two concepts mainly involve the electromagnetic nature of particles and atoms. It may be intriguing and helpful for readers to understand the history behind these concepts.

1.4.1 Checkpoint up to this part of the chapter

<u>Electricity and Magnetism concepts</u> we will next be 'dissecting' to help you understand more concepts concerning MRI.

a) $\Delta E \rightarrow M$ and $\Delta M \rightarrow E$ are thematic in the understanding of MRI, especially in terms of excitation and reception of signals. These concepts are essential building blocks for chapter 2, where we learn about the MRI Hardware.

b) MRI relies on imaging that results from magnetic dipoles and precession of these dipoles. The precession frequency is known as the Larmor frequency in MRI. We will see that dipoles are the fundamental concept for chapter 2, when we discuss the main magnetic field and the generation of signals.[23]

However, at this point, let us take a moment to refresh our knowledge concerning electricity and magnetism (E&M).

1.4.2 Review of basics of electricity and a little bit of history

The concept of electrostatics, which you may have already learned about in high school, encompasses stationary electric charges. These E&M charges are important to the understanding of MRI, as charges are a fundamental property of the universe.

Figure 1.29. Checklist of core concepts already completed at this point of the chapter.

---

[23] Additionally, both dipoles and precession are essential to the understanding of T1 and T2 relaxation, which are described in chapter 3.

The word electricity is derived from the Greek word 'elektron', the Hellenistic[24] name for amber, which is a gemstone. It has been thought that in ancient Greece, people noticed that amber would attract feathers, straws, and other light materials when rubbed with a cloth or fur. Fast forward to later days where a much deeper understanding of moving charges began to evolve. This was notably investigated starting in the 18th Century, particularly due to some of the initial contributions of Benjamin Franklin, who demonstrated the effects of electricity from lightning, as seen in figure 1.30. During that time, Franklin and his

Figure 1.30. Benjamin Franklin depicted holding a key attached to a string and a kite that has been struck by lightning on a U.S. postage stamp, which marks some of his early investigations in the subject of electricity. Reproduced from Blue Moon/stock.adobe.com.

---

[24] The word 'Hellenistic' refers to the cultural and linguistic influences of Greek culture on the ancient world at that time.

colleagues investigated the effects of electricity and even knew that electricity passed through metals. Later in this chapter, we will also discuss more about the discoveries by Michael Faraday and James C Maxwell, who made additional contributions to E&M in the 19th Century. These discoveries are further discussed in section 1.5.

To create a greater personal understanding of E&M, let us look at things around us that have static charge. Note in figure 1.31 how static charges can attract polar water. You can see this effect directly in your own sink! However, we can also see this effect using our own bodies.

To create a greater personal understanding of E&M, let us look at things around us that have static charge. Note in figure 1.31 how static charges can attract polar water. You can see this effect directly in your own sink! But, we can also see this

Figure 1.31. The balloon in this figure illustrates how static charges on its surface attract the polar water molecules toward it. Note how the static charge of the balloon alters the pathway of the flowing water. Reproduced from harunyigit/stock.adobe.com.

Figure 1.32. Static hair is the result of electrical charges being transferred to hair and exhibiting a repelling effect. Perhaps this sometimes happens to you between your hat and your hair [28]. Adobe Stock Images: © RachelKolokoffHopper

effect using our own bodies. Did you ever wonder what causes static in your hair, as in figure 1.32?

Static electricity is created when two objects rub against each other and transfer the charges. Without any humidity or moisture in your hair, like on a dry winter morning, the charge causes the strands to repel from another like a magnet [29]. These day-to-day examples remind us of static charge on a personal basis. Perhaps, this brief review may have you recalling static charges.

1.4.3 Electricity and magnetism and the world around us

We will start by describing the world of MRI through a quick reflection on electricity. You might have encountered the concept of charge particles, such as shown in figure 1.33. This topic was taught pervasively not only in your physics classes, but also in your chemistry classes. Moving charges also include basic scientific concepts with which we are already familiar. Electricity surrounds us in our daily environment, as well as in Nature. Electricity powers our laptops, as shown in figure 1.34. As we plug our computer into the wall socket, we can begin to visualize the flow of charges that comes from the wall socket and enters and powers our computer. In Nature, we consider the electrical discharge of lightning. Lightning is much more dramatic than plugging in a laptop; however, it is fueled by the same distribution of positive and negative charges. The moment before electrical discharge happens, positive and negative charges align on two sides. Between them is something we will learn about next: the electric field. First, though, we will state our learning objectives.

Figure 1.33. The concept of a charge has been an 'iconic' tool with which we are quite familiar in our chemistry and physics classes. Reproduced from benjaminec/stock.adobe.com.

Figure 1.34. We use electricity every day. Electrons power our laptops and enable us to access information from around the world. There are also examples found in nature; one that comes to mind readily is lightning through electric discharge.

Amy B. (psychology graduate student) reflected on Electrostatics and E&M charges:

"Thanks for this reminder about the basics of static electricity. It helped me see the Electricity and Magnetism (E&M) foundational concept that similar electrons attract and dissimilar electrons repel each other."

1.4.4 Learning objectives

Proceeding forward, we will discuss the essential learning objectives for this section. These objectives will help us understand changing magnetic and electric fields and how they create the events that make MRI possible.

(1) Review the definition of a field (physics).

(2) Describe what a magnetic field is—i.e., a spatial area in which magnetic or electric particles will experience a 'force.'

(3) Discuss how current in a loop can produce a magnetic field.

(4) Understand how a magnetic field in a loop induces a current.

(5) Evaluate an example of a real-world device that relies on one of the principles ΔE field $\rightarrow M$ and $\Delta M \rightarrow E$ field (Note: current flows because something is producing an electric field that forces the charges around the wire, as can be seen in figure 1.34).

1.4.5 Michael Faraday and the EM field

In order to understand EM waves, we must understand the EM field. Note that an EM field is made up of both electrical and magnetic components [10].

Electromagnetism is known as a 'fundamental force' in physics [30]. Another example of a fundamental force is the gravitational force. You are probably well aware of how the Earth's gravitational force works on us each day.

In this section, we will try to unravel one of the most fundamental sets of laws in electricity and magnetism. Please find and watch the video entitled 'Einstein's Big Ideas.' I'd highly recommend you watch it, as it has the drama that surrounds the amazing discovery by Michael Faraday [31].

In physics, a field means that a physical quantity is assigned to every point in space (or, more generally, spacetime). A field is seen as extending throughout a large region of space so that it influences everything—'an area.' The strength of a field usually varies over a region. Michael Faraday (figure 1.35) was thought to be the first to coin the term 'field' [32].

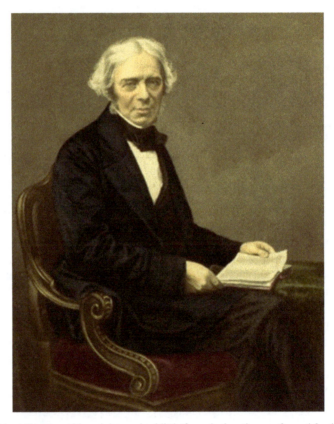

Figure 1.35. Michael Faraday, although he received little formal education, performed fundamental research on magnetism. It was Faraday who established the basis and initial concept of the electromagnetic field in physics. Reproduced from Georgios Kollidas/stock.adobe.com.

Natalie (pre-med biomedical engineering student) reflects on fields:
"When I hear the word field, the first thing that comes to my mind is like a softball field (I played shortstop in school). I do not think this is what Faraday had in mind when he coined the term, but that fact that every point in space has a response (like a force) to items that enter that field.

Also, what immediately comes to mind is a magnetic field, and I know we have all seen that in pre-med physics. This is where the instructor showed that we have the negative and positive ends of a magnet with arrows extending out from each end, showing the field (this can be seen below in figure 1.38). A more formal way to think of a field is something that extends through a large region of space that interacts with physics. In the physics realm, though, this field influences everything around it and has a strength that varies, depending on what point in the field you are at."

Dr. Wu responded:
"Sounds like you are on your way to understanding fields in the context of E&M."

Amazingly, the use of 'field' appears for the first time in Faraday's paper, 'On new magnetic actions,' in 1846. Faraday was referring to a region in the vicinity of a magnet affected by physical forces. Here is a direct quote from Faraday describing a 'field' [32]:

'the capability of placing magnetic portions of matter one within another, and so observing dynamic and other phenomena within magnetic media. In fact, not only may these substances be placed as magnets in the magnetic field, but the field generally may be filled with them, and then other bodies and other magnets examined as to their joint or separate actions in it.'

1.4.5.1 Scaffolding box: review of vectors and vectorial pictures
Here we provide a very brief review of a vector and vector fields. A vector quantity represents direction and magnitude, depicted as an arrow. Vectors are commonly used in physics to represent component and net force within a system. See vector representations in figure 1.36.

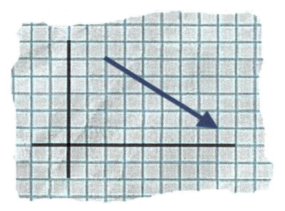

Figure 1.36. The vector above represents a net displacement of 6 units to the left and 4 down, written as (6, -4).

Figure 1.37. You may also have encountered or find useful vector fields. Note the vector field showing wind shear in the map above. This image has been obtained by the author from the Pixabay website where it was made available under the Pixabay License. It is included within this article on that basis.

A magnetic field is a vector field that describes the magnetic influence of electric charges in relative motion and magnetized materials. You have previously seen other kinds of vector fields, probably including ones like the wind current map shown in figure 1.36. You may also have encountered or find useful vector fields. Note the vector field showing wind shear in the map. At each point in space, a certain physical force can be visualized, as demonstrated in that image. Similarly, a magnetic field can also be described as the magnetic effect of electric currents and magnetic materials and expressed as a vector on a map (figure 1.37). The magnetic field at any given point is specified by both a direction and a magnitude (or strength).

It is helpful recognize this pattern of a vector field. These vector fields are common to the depiction of magnetic fields, as seen as in figure 1.38.

Figure 1.38. Note on the (left) iron filings follow the force lines around a simple bar magnet (reproduced from milkchocolate/stock.adobe.com). (right) These can be turned in a visual vector field, which would also indicate the directions of the flow (reproduced from Vasily Merkushev/stock.adobe.com).

1.4.5.2 Historical reference bubble: James Clerk Maxwell: consolidation of electrical and magnetic phenomena through his famous equations still used to this day

James Clerk Maxwell (shown in figure 1.39) studied and commented on electricity and magnetism in his paper 'On Faraday's lines of force.' Maxwell presented a model of Faraday's work and how electricity and magnetism were related using a set of differential equations. We still apply Maxwell equations today in MRI and other E&M devices. Maxwell also calculated that the speed of propagation of an electromagnetic field is that of the speed of light. He showed that the equations could predict the existence of waves of oscillating electric and magnetic fields that travel through space at a speed that were consistent with results as predicted from simple electrical experiments [9].

Figure 1.39. James C Maxwell was a Scottish scientist whose work unified electromagnetic radiation equations. In this work electricity, magnetism and light were into a unified set of physics equations. This 'File:James Clerk Maxwell.png' image has been obtained by the author from the Wikimedia website, where it is stated to have been released into the public domain. It is included within this article on that basis.

Summary:
1. Static charge and attraction and repulsion of charge.
2. Fields.
3. There are equations (Maxwell's equations shown in section 1.4.5.2) that can be used to describe the physics of the EM field.

Gail H. (neuroscience postdoctoral researcher) reflects on E&M concepts:

"From reading these sections, it seems like Faraday was the great experimentalist who demonstrated a lot of evidence that led to these ideas. It was then I guess Maxwell who wrote them down as coherent equations that we use even today. I think when I watched the videos it was really super interesting. Some of the interesting terms that Maxwell figured out in the E&M equations actually correlated with the speed of light, making us realize that light itself is an E&M phenomenon. Light is very fascinating and I'm really happy that I learned this."

"In physics, electromagnetics consists of waves of the electromagnetic (EM) field, propagating through space, that carry energy."

1.5 Changing fields alter E&M

1.5.1 Concept number 4, changing electric field → makes magnetic field, and changing magnetic field → makes electric field, which we can also provide names to as Faraday's law ($\Delta M \to E$) and Ampère's law ($\Delta E \to M$)

Thematically, these changes operate in the direction that follows the **right-hand rule** that is often taught in high school and university physics classes. The right-hand rule is a memorable tool that we have found helps connect students to the concepts of $\Delta E \to M$ field and $\Delta M \to E$ field that we would like students to review and gain greater intuition. To provide a name for these terms, we will describe them as **Faraday's law** ($\Delta M \to E$) and **Ampère's law** ($\Delta E \to M$) at this point, but we will next try to provide more physical intuition about these concepts (as pictorialized in figure 1.40) by suggesting an experimental lecture setup and pointing to a couple of videos on the topic. Note, as mentioned in the historical box, that it was James C Maxwell who combined the insights by Faraday and Ampère together mathematically that helped consolidate the concepts (see section 1.4.5.2). These Maxwell equations are in common use by engineers and physicists to this day, and this topic is highly relevant to MRI scanner understanding and design.

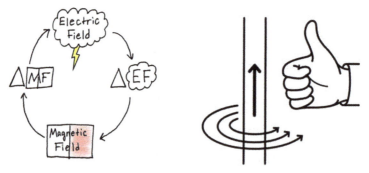

Figure 1.40. (left) The central idea shown as a conceptual sketch of the relationship between electrical and magnetic fields as the 'key' point of this subsection 1.5 (right). Also please review the right-hand rule, as the direction of changing field effects is more easily recalled by using this simple hand mnemonic.

1.5.1.1 Applied bubble: a simple thought experiment and application

If you are not able to perform this experiment that is seen in figure 1.41, then it is informative to look at an image and imagine the effect of moving a magnet bar between a loop of coil as shown in that figure. One way to gain more intuition on the laws of E&M is to see that if you move a magnet back and

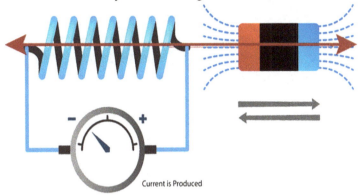

Figure 1.41. One way to gain more intuition on the laws of E&M is to see that if you move a magnet back and forth through a loop of coil, a current will be created through the attached wire, demonstrating that a changing magnetic field can produce a current. Reproduced from VectorMine/stock.adobe.com.

forth through a loop of coil, a current will be created through the attached wire, demonstrating that a changing magnetic field can produce a current. Do not worry; we provided another experiment with a motor and voltmeter that also provides physical intuition. However, the reader might consider watching the following Khan Academy [33] video that provides additional information on the topic.

1.5.2 Putting together $\Delta M \rightarrow E$ and $\Delta E \rightarrow M$—suggested experiment to run in class

As seen in figure 1.41, a **central idea** that we will discuss throughout the book is that a changing magnetic field in a loop can induce a current, as well as how a changing electric field in a loop can create a magnetic field. This can feel somewhat abstract, and if it does, you are not alone! Realize that the discovery of this effect did not even arise until the 19th century under scientists such as Faraday, Ampère, Maxwell, Gauss, and others who were able to understand this concept.

Below is a simple experiment that you can do (or have your instructor do), with items that are relatively inexpensive ($25–30 as of the current publication of this book). This experiment demonstrates how a magnetic field can induce a current. Seeing is believing!

Items shown in figure 1.42 are used for an experiment to demonstrate the effect of changing a magnetic field inducing a current. The wrench shown is just for convenience. You can perform the experiment without vice grips or a wrench, as long as you have a way to grip the ends of the motor so you can physically turn it.

For further description, see appendix 1.A.1 for the steps we used to help set up the experiment.

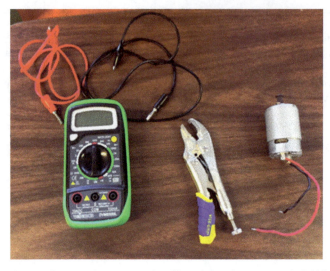

Figure 1.42. Items for experiment to demonstrate the effect of changing a magnetic field inducing a current. The wrench shown is just for convenience. You can do the experiment without it.

 Nick P. (third-year medical student) reflects on $\Delta M \rightarrow E$ and $\Delta E \rightarrow M$:
"So, I understand that MRI scanner uses current passing through coils to create the magnetic field, but what role does the magnetic field play in producing the images that we are able to see?"

 Dr. Wu replied:
"Hey Nick, that is a great question that lets me preview what is coming up in chapter 2, where we discuss the MRI hardware through how the machinery works. In chapter 3, we mainly discuss relaxation, signal-to-noise, and contrast. Water content plays some role in the signal, but what you end up seeing will be described by image weightings, which can be described through the lens of relaxation (decay) that creates the image contrast that our radiologists look at. I think your question about how magnetic fields play to produce images is a sophisticated one and will require the reading of several chapters to get to the 'meat' of the issue. Keep reading on, I think you are grasping these concepts."

1.5.3 Electromagnets

The main magnet (which we will learn more about in chapter 2) used in a majority of clinical magnets are of the electromagnet form. At this point, it is instructive to see how a magnet can be constructed from electric current and coil loops. An electromagnet is a magnet that runs on the principle of electricity, as shown in figure 1.43. We are able to create a changing electric field through the use of coils because the

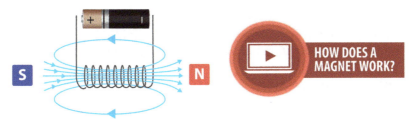

Figure 1.43. A video is presented that illustrates how to make a simple electromagnet [34, 35]. Reproduced from Fridas/stock.adobe.com.

current is changing directions in a loop. Note the interesting fact that unlike permanent magnets (discussed in the appendix), the strength of an electromagnet can be altered by changing the amount of electric current that flows through it and/or just by increasing the number of loops. We will revisit this concept in chapter 2 when we discuss the main magnetic field. Another thought-provoking fact is that we can also simply reverse the north and south poles on the electromagnet just by changing the direction of the flow of electricity.[25]

1.5.3.1 Physical laws

An electric circuit and magnetic field interact and can be 'transformed' into the opposite, via Faraday's law ($\Delta M \rightarrow E$) and Ampère's law ($\Delta E \rightarrow M$). Note that Maxwell equations are how most physicists have integrated the concepts of E&M together neatly into mathematical rules that include additional concepts, but we have simplified the concepts here for the reader.

Discovered in 1823, Ampère's law [36] relates the current passing through the loop to be equal to the 'integrated' magnetic field. A current would be generated by a changing electric field, for example. Another way of saying it is that the law states that for any closed loop path, the sum of the length elements times the magnetic field is equal to the electric current.

Faraday's law states that a change in the magnetic environment of a coil of wire will cause a voltage (EMF) to be 'induced' in the coil. Faraday's law of induction is one of the important concepts of electricity. It looks at the way changing magnetic fields can cause current to flow in wires. Basically, it is a formula/concept that describes how potential difference (voltage difference) is created and how much is generated. It's a huge concept to understand that the changing of a magnetic field can create voltage [10]. Faraday's experiments showed induction between coils of wire. In his experiments, the liquid battery provides a current that flows through the small coil, creating a magnetic field. When the coils are stationary, no current is induced. But, when the small coil is moved in or out of the large coil, the magnetic flux through the large coil changes, inducing a current that is detected by the galvanometer [34].

---

[25] In appendix 1.A.3.2, we describe a different type of magnet (the permanent magnet). In that case, the magnet has been exposed to a field and no longer needs the changing electric field (i.e., current) to preserve its field. In the case of the electromagnet, which is the majority of MRI systems today, if the magnet loses power, then the system will quickly lose field.

Finally, we move on to the very last concept, which is dipoles and precession. With this last core item, we believe you can now scaffold a big chunk of MRI and start focusing on the more intricate and intriguing parts of MRI.

1.6 Dipoles and precession

Despite the complexity of the topic, we will provide a conceptual approach to understanding E&M. For those who are curious, please refer to the end of our chapter for several references that provide more physics derivation. For the purpose of a conceptual understanding, we provide you with the following learner goals that are focused on the practicalities that an applied clinician or researcher may consider for first understanding.

Learning objectives:
1. Dipoles model two physically separate 'poles' of opposite force. For magnetism, one can generally consider the north and south poles of a magnet.
2. Dipole is a vector consisting of both direction and quantity.
3. Examine the difference between electrical dipoles and magnetic dipoles.
4. Understand that we are concerned with magnetic dipoles, which are equivalent to a theoretical infinitesimal ring of current.
5. Conceptualize the idea of precession.
6. What is mechanical angular moment?
7. Elements needed for precession (i.e., tilt in the angular momentum and gravity/main magnetic field).

As we discussed earlier, water functions as the key element to be measured in MRI. Water is similar to a magnet, and the following section will describe how water is a magnetic dipole. A central motivation for understanding the dipole is that it forms the signal carriers in MRI. In chapter 2, the concept of measuring water molecules will be further explored as it relates to hardware. The signals produced by water molecules are very small in scale. Therefore, it makes sense that we would need a large magnet such as that we use for MR scanners to be able to measure those signals. In order to understand how this process works, we will proceed with reviewing what is a dipole.

1.6.1 What is a dipole?

Breaking down the concept, let us start with the basic word '**di-pole**.' The word DI-POLE indicates two poles, i.e., having two opposites. The type of dipole that we are interested in is two opposing charges that are separated by a distance, such as an electric dipole or, more importantly, a magnetic dipole. The magnetic dipole is the source of the magnetic moment in MRI.

How do we study the dipole? First, take a positive charge and then put a negative charge some distance away. The easiest way to think about these charges is with a charge $+Q$ (where Q is a numerical value) and a charge $-Q$. There are several reasons that cause these molecules to be charged. One is molecular structures: there are positive charges in one location and negative charges in another. One particular location of interest is the outside orbital of elements that are involved in the construction of a molecule, as you may have previously learned in your chemistry classes.

This can be seen in molecules such as NaCl (table salt), where the chlorine in one part of the molecule is electronegative and the sodium in another part of the molecule is more positive. The presence of opposite charges results in polarity that acts like a vector. A vector is a quantity that has a direction. With a positive and a negative charge, there is a distance between the two that creates the dipolar model. When you stand very far away from it, all you see in the distance is a single molecule. Up close, you can see the individual charges. For the most part, if you start to add up many of these dipoles as vectors, you can begin to see a magnitude and a direction in terms of a population of dipoles. In MRI, we think of the population of magnetic dipoles as forming something called an isochromat [38, p54], or the sum of all the individual dipoles as a unit.

We wanted to trigger your memory from your previous chemistry classes (chemistry is a prerequisite class for medical students, radiological technologists, and many other scientists and engineers). If you recall the ionic bond, it may help further visualize the dipole. As seen in figure 1.44. NaCl, an ionic compound, is shown with chemical bonds. A concrete way to think of practical dipoles is to consider table salt, or in chemistry nomenclature, NaCl, written with its charges as Na^+ and Cl^-. The ionic bonds between sodium and chlorine are shown with orbital models. As will be seen in the figure below, the current is in an infinitesimal current loop with a magnetic dipole.[26]

1.6.2 Magnetic dipole

It is important to understand that we are concerned with magnetic dipoles, which are equivalent to a theoretical infinitesimal current ring. Magnetic monopoles (i.e., isolated magnetic charges) don't exist in Nature. Isolated 'm' charges don't exist as single elements, but according to our current physics knowledge, we believe that magnetic charges +m and –m come in pairs (i.e., magnetic monopoles are currently considered not to exist in Nature).

Understand that a dipole is two poles. It represents or models the effect of two physically separated equal and opposite charges. For magnetism, consider the N and S poles of a magnet as representing opposite 'poles' of a system, as illustrated in figure 1.45.

Another way to create a magnetic dipole is to use electricity as a current in a ring. Remember that we can turn a changing electric field into a magnetic field. So if you can imagine a small infinitesimal loop of coil that had a small current in it, the resulting system would function as a magnetic dipole.[27] It is useful to

---

[26] Note that we can also apply the $\Delta E \to M$ and $\Delta M \to E$ concept for relating to this concept for the infinitesimal coil loop [38].

[27] Brown *et al* describe the concept of this model in their book [38]. Recall a magnetic dipole moment as a pair of magnetic charges with equal magnitudes and opposite signs separated by distance. The choice of the word 'dipole' is analogous to the way electric charges lead to electric dipole moments, though the concept for magnetic dipole perhaps is better represented as the current loop. We will describe more about this concept in section in 1.A.3.1 when it is also referred to as the magnetic moment. The word 'dipole' has its origin in this model, which is acceptable to the current loop picture as long as we use it to investigate only the field outside the moment structure.

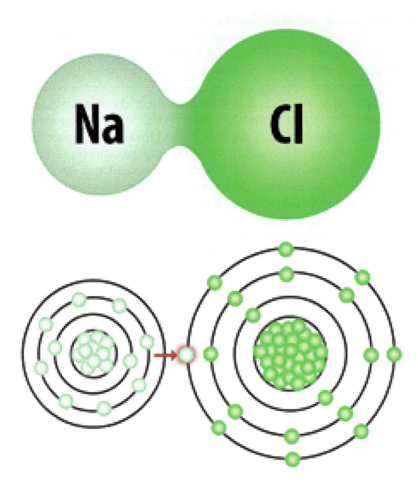

Figure 1.44. NaCl, an ionic compound, is shown with chemical bonds. A concrete way to think of practical dipoles is to consider table salt, or in chemistry nomenclature, NaCl, written with its charges as Na$^+$ and Cl$^-$. Reproduced from designua/stock.adobe.com

understand this model for future purposes. The direction of current proceeds in a loop similar to how a magnetic field can generate current in a wire, as shown in figure 1.46.

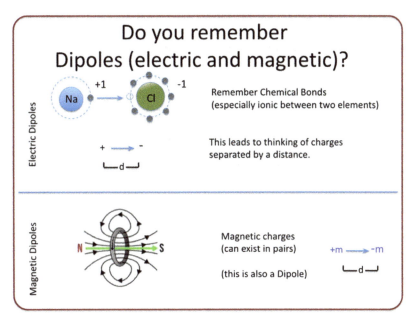

Figure 1.45. In this figure, we show applications of an electronic dipole (top), and a representation of the magnetic dipole (bottom) [39].

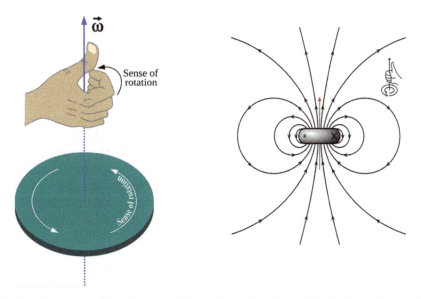

Figure 1.46. A visual representation of a magnetic dipole (shown in red) as modeled by the small loop coil. Note, the direction of the moment follows the right-hand rule, as shown in the figure to the left (reproduced from ScientificStock/stock.adobe.com). A basic review of the right-hand rule (RHR) can be seen at Khan Academy. Physical processes. MCAT. Test prep [40]. This Dipole magnetic-ring.svg image (right) has been obtained by the authors from the Wikimedia website where it was made available by User:MikeRun under a CC BY-SA 3.0 licence. It is included within this article on that basis. It is attributed to User:Geek3 and User:MikeRun.

Carlie P. (dancer and premedical student with chemistry and psychology emphasis) reflection on the magnetic dipole:
"It seems like a bit of leap to go between the electric dipole and the magnetic dipole. If there are isolated electric charges (monopoles) that when bonded together form a dipole, wouldn't that suggest there are magnetic monopoles too? It's hard for me to imagine why a magnetic monopole does not exist, but I better understand this concept when imagining a ring. While I can identify points along the ring that are on opposite sides, it is an infinite loop that does not have individual parts."

Once we understand the 'magnetic' dipole, we can move on to deeper concepts such as precession, which is the final part of this section and fifth core concept of this chapter.

Natalie N. (premed) said:
"It's nice to see this concept gone over several times—it's making it more clear. Also, it seems like there are several terms that mean about the same thing, which is the magnetic dipole/magnetic moment, which are actually two ways to refer to a very similar principle in physics. I'm looking forward to seeing what happens next, but I can also visualize this vector as being something that we have to worry about or think about."

1.6.3 Precession

We will find out in chapter 2 that the MRI scanners operate and work at the precessional frequency (known as the Larmor frequency in MRI). The precessional frequency is an important concept in the hardware of MRI. In fact, each magnet will have to be tuned (much like a tuning fork to a certain frequency of operation) to a specific frequency, as described by Catherine Westbrook [41]. To be useful, there will be a specific frequency range for each field strength and molecule we are attempting to image. In the clinical MRI case, this is typically water. First, it is important to take a step back and examine what is precessional motion in simple mechanics.

We found that it was most intuitive for students who may not have studied lots of physics or have not seen it in many years to just visualize the spinning gyroscopes and understand that there has to be intrinsic spin in protons for those spins to be tipped and for the entire system to be placed in a large field (such as a gravitational field for mechanical gyroscopes, and a magnetic field for protons in MRI). For those who are interested in the physics approach to learning the topic, please see appendix 1.A.3. Please consider watching the video link on precession in figure 1.47.

Figure 1.47. Gyroscopic motion is a key element to understand when it comes to the physics of precession. If you like animations and want to get additional animation to help you see the vectors, refer to the video. This illustrates how the angular momentum vector is affected by torque, causing precession. Note that appendix 1.A.3 further delves into the physics of precession. For those who are interested in more information, see [11].

The spinning motion of particles is known as precession.[28] At any moment in time, all of the billions of hydrogen protons in our bodies are in random positions and spinning on their own axes. Water is made up of two hydrogen atoms and one oxygen atom. The hydrogen nucleus contains one positive charge a proton spinning around on its axis, which acts like a tiny magnet reordered as an animation in figure 1.48. Spinning objects, including protons, have angular momentum. Angular momentum points along the axis of rotation. For those who are interested in the physics approach to learning the topic, please see appendix 1.A.4. Please consider watching the video link on precession in figure 1.47. Gyroscopic motion is *a key element to understand when it comes to the physics of precession. If you like animations and want to get additional visualizations to help you see the vectors, refer to the video. This illustrates how the angular momentum vector is affected by torque, causing precession. Note that appendix 1.A.4 further delves into the physics of precession, for those who are interested in more information* [11].

We will find out in chapter 2 that the MRI scanners operate and work at the precessional frequency (which is referred to as the Larmor frequency in MRI). We will discuss Larmor frequency in a little more detail in chapter 2.

---

[28] All physical materials have spin. One form of spin is a quantum-mechanical property akin to angular momentum in classical physics. Spin angular momentum is an intrinsic (built-in) property, just like rest mass and charge. Spin is carried by elementary particles and atomic nuclei. In MRI, we are most interested in the proton, which we depicted in figure 1.22 (the spinning). A more detailed chemistry-based description is contained in [42].

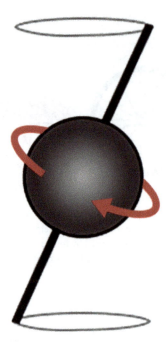

Figure 1.48. A simplified look at precession. Notice the axis of the spin. Animation available at https://doi.org/10.1088/978-0-7503-1284-4.

1.6.4 Putting together the main idea: protons have 'spin' and thus angular momentum, so they have ability for precession in the electrical sense just like gyroscopes have the ability to precess in the mechanical domain

The Greek symbol phi shows the tipping angle on the picture, which refers to the angle relative axis of precession (which is indicating what we will come to find out will be the main field).[29] The tipping angle will be later known as the 'flip angle.'[30]

 Justin N. (radiology attending) reflects on gyroscopes:
"I think the gyroscopes are very helpful in getting a truer concept of what is happening on a molecular level in a magnetic field. Just being able to see that not every molecule is perfectly lined up and may in fact be upside down is helpful."

Table 1.1 provides an important summary of precession for MRI. If you memorize it and have gained some intuition of what the concept of precession is

---

[29] We will learn more about the main field in section 2.2, as it is in the direction of the cylindrical bore of the magnet.
[30] We will learn about the RF transmitter, which will be responsible for tipping the spins off the main axis, in section 2.5.

Table 1.1. Comparison of roles of the mechanical gyroscope and the analogous particles in the E&M domain (the latter relevant for MRI).

Mass	MRI/NMR proton
Spinning top	Spin in the 'proton'
Angular momentum from mass rotating	Angular momentum intrinsic to atomic particles (the proton), intrinsic magnetic moment
Gravitation field	Magnetic field (main field) must be very strong to get the number of spins to be converted to signal
Tilting top, I tipped manually mechanically in this case	Tilting magnetic moment. But what is going to be able to tip it? (See RF transmit later)

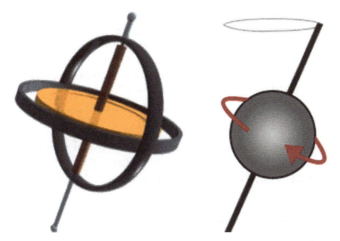

Figure 1.49. Gyroscope example of physical precession compared with a proton example of atomic precession. This 'File:Gyroscope precession.gif' image has been obtained by the author from the Wikimedia website, where it is stated to have been released into the public domain. It is included within this article on that basis. Animation available at https://doi.org/10.1088/978-0-7503-1284-4.

and are able to relate it to spinning protons, it will help you in future concepts through MRI, as seen in figure 1.49. It is not entirely essential that you fully grasp the derivation. It may be sufficient for you to conceptualize that there is the effect of E&M in MRI interaction with the protons in the body and that there is precession. However, some of you may find that the topic is demystified to a point that you are able to apply the concepts to connect the dots.

The Greek symbol phi (ϕ) shows the tipping angle on the picture, which refers to the angle relative axis of precession (which is indicating what we will come to find out will be the main field).[31] The tipping angle will be later known as the 'flip angle.'

---

[31] We will learn more about the main field in section 2.2, as it is in the direction of the cylindrical bore of the magnet.

Table 1.1 provides an important summary of precession for MRI. If you memorize it and have gained some intuition of what the concept of precession is and are able to relate it to spinning protons, it will help you in future concepts through MRI, as seen in figure 1.49. It is not entirely essential that you fully grasp the derivation. It may be sufficient for you to conceptualize that there is the effect of E&M in MRI interaction with the protons in the body and that there is precession. However, some of you may find that the topic is demystified to a point that you are able to trace the motion of **precession**.

This understanding will aid your ability to comprehend the clinical environment, which includes 'relaxation,' MRI hardware, and pulse sequences, as well as advanced applications later. As you work through those ideas, keep the concept of precession and dipoles in mind.

Nick P. reflects on what is different or similar for precession in the gyroscope versus precession of protons in a magnetic field:
"Table 1.1 helps solidify the different components that go into the generation of the precession of the protons, and really helps me get a better image of what is going on."

If you have not quite yet understood all the concepts of spin and precession, you will have further chances to see these concepts again in chapters 2 and 3, where we will apply these ideas directly to MRI hardware and relaxation.

1.6.5 Knowledge check for the magnetic dipole

1. What is a dipole?
2. Can you describe the difference between electric dipoles and magnetic dipoles?
3. Does water have both electric and magnetic dipoles?
4. Can you describe the elements that are required for precession?

Answers:
(1) A dipole represents two separate poles (one positive, one negative). The magnitude and direction of the charge separation is indicated by using an arrow (typically drawn starting from the positive charge, then pointing to the negative direction).
(2) An electric dipole can be visualized as separate but opposite charges. However, we visualize the magnetic moment simply as a dipole arrow and/or a circular ring of current (they have a similar function, except that magnetic solitary charges do not exist).
(3) Yes, water has both electric and magnetic dipoles.
(4) For mechanical precession, you need spinning angular momentum, in a gravitational field with mass, something to tip the 'top' (torque is a turning

Figure 1.50. (Left) is a micro motor (silver). It functions through conversion in E&M energy. It has two leads (red and black) that can be attached to the multimeter to convert magnetic current to electrical current [35]. (Right) inside the motor are magnets. The spinning assembly interacts and changes the fields (i.e., ΔM), so this is what is generating the current.

force with lever arm). A top and a proton can be used to portray precession, as shown in figure 1.50. For greater details, see appendix 1.A.3.

I-A Appendix of chapter 1

Five topics are included in this appendix:

In section 1.A.1, you will learn details about the less-than-$25 experiment that demonstrates that a changing magnetic field produces current and a changing current can generate a change in magnetic field. We found this to be the easiest way for students to gain some intuition to see the direct application and observation of effects in class. Note, Faraday's law ($\Delta M \rightarrow E$) and Ampère's law ($\Delta E \rightarrow M$) are relied on, particularly in chapter 2, when we cover how the RF receiver and RF transmitters work.

In section 1.A.2, you will learn scales of magnetic fields, units, and relative field strength as compared with the earth's magnetic field. Learning the units of gauss and Tesla and their relative scale to the earth's magnetic field can provide intuition about the strength of the fields produced in MRI scanners.

In section 1.A.3, you will receive a brief description of the physics of precession and change in angular momentum. The intrinsic knowledge of the mechanism of precession is obtainable by an understanding of torque, and changing angular momentum in a mechanical model can aid in developing intuition about the physics of precession. Some advanced readers will be able to put together the picture that incorporates application of the right-hand rule.

In section 1.A.4 you will conceptualize alignment of spins in a strong magnetic field. Parallel and antiparallel spins will again be revisited in greater detail in chapter 2. This parallel and antiparallel concept of spin are essential to the understanding of signal-to-noise levels in MRI also covered in that chapter.[32]

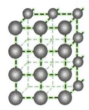
In section 1.A.5 you will understand what the 'lattice' in MRI is (the topic is helpful to preview this concept for chapter 3, when we cover T1 relaxation in more detail). Some students may have heard of MRI relaxation (T1 and T2, for example). Relaxation forms the key element for understanding of image contrast, and spin-lattice requires the understanding of what is a 'lattice.' Thus, this appendix can preview this fundamental topic before we explore the idea in later chapters in applications.

1.A.1 Description of the experimental setup for $\Delta E \rightarrow M$ and $\Delta M \rightarrow E$

Want to try your own E&M experiment for under approximately $25 at the time of writing this book? Also, you might be able to borrow the multimeter, and a small motor could be found at any university lab, extracted from a discarded toy, or ordered easily online, which could reduce your cost. Below are a few links that we used at the time of writing this book, but you may have to search for them if the links are no longer working at a later date.
  1. Digital multimeter [43].
  2. Motor. You can see the addition of the electric current [44].

Note that vice grips (see vice grips in figure 1.50) provide easier manipulation of the motor. Alternatively, you can use any wrench or attachment device, and even some paper clips could be bent to create a lever to turn the motor.

Below is a digital multimeter, which measures electrical current and voltage. A digital multimeter is a test tool used to measure two or more electrical values—principally voltage (volts), current (amps), and/or resistance (ohms). It is a standard diagnostic tool for technicians in the electrical/electronic industries (figure 1.50).
1. Attach the positive and negative ends of the leads to the respective spots on the multimeter (see figure 1.51).
2. After connecting, turn on the multimeter to any voltage. It will read at 0 or –0. In order for the multimeter to read a voltage, you must create a current. A wrench

---

[32] This is important, as when working with equipment you will need to understand what parameters are being used by that hardware, not only for day-to-day operations, but also for when you are purchasing equipment for your facility. It is also essential to understand parameters for MRI safety, as certain devices are conditional for scanning in the MRI environment. Parameters will allow you to meet guidelines. Without that knowledge, it will be more difficult to ascertain the appropriateness of conditions and the components used in MRI.

Figure 1.51. Step 1—attach leads with alligator clips. Note that these motors do not necessarily require polarity, so you connect the black and red to either side of the motor.

attached to the motor will do this. However, as you can see, when just attached, the meter still reads at zero (see figure 1.52) before you have rotated the wrench.
3. Then, once you begin to rotate the wrench around the motor, the multimeter detects a current due to the changes in the magnetic field inside the motor (see figure 1.53 to observe a magnet configuration inside of the motor). Rotating the wrench causes a disturbance in the charges of the magnetic field. Then, these charges create a current that can be transformed into electrical activity, since magnetic fields are caused by electric currents.

Finally, recall the times you've put batteries in a toy and suddenly you see a motor turning, such as in a toy car. This comes from the fact that a changing current can induce changes in the magnetic field. The magnetic field drives the motor of the car! I hope if you do this experiment that you also feel the significance of the motor and what it has meant to us in our lives, as many things would not be possible without the electric motor.

1.A.2 Units

The familiarity and sense of the scale of units and the magnitude of magnetic units is important to MRI (figure 1.47). Magnet units are a licensing board testable topic and can be equally important to learn for clinical understanding.

Below is a bulleted list for you to memorize:
- The goal of this section is to understand scales of magnetic units (Tesla and Gauss) and relative strength to as it relates to the strength of Earth's magnetic field.

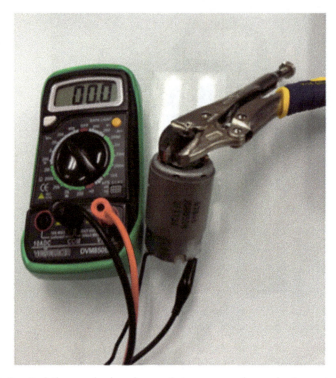

Figure 1.52. Step 2—attach the wrench, vice grips, or whatever you can form a simple way to grip the motor so you can turn it quickly.

Figure 1.53. Step 3—rotate the wrench or lever you created. You will see that the multimeter registers a voltage indicating that there is current being generated by the action of the changes in the magnetic fields of the motor.

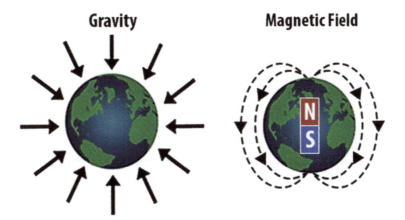

Figure 1.54. As shown on the figure to the left, we are well familiar with how gravity influences the space around the earth is used to explain the influence that a mass extends into the space that surrounds that area above the earth's source. This gravity is producing a force on other bodies of mass. However, there is also a magnetic field, as shown on the right figure, in the Earth's field.

- An MRI magnet has 20,000 times the Earth's magnetic field. As you learn more about MRI, it is important to learn the scales of units of the magnetic field. 1 Tesla = 10 000 Gausss.
- Note, for a sense of scale of field strength, the Earth's magnetic field is about 0.5 to 1.0 Gausss.
- Main magnetic fields are on the order of Tesla (i.e., 1.5 T or 3 T) and the gradient strengths are on the order of tens of Gausss cm^{-1}.

1.A.3 Further discussion on magnetic moments and magnetic domains

This section follows up on key concepts related to the magnetic moment. As we previously discussed in the dipole section, we have a north and south pole associated with magnetic dipoles. The missing link to these concepts is that these poles are separated by a distance. A moment is a physics quantity. In physics, a moment is an expression involving the product of a distance and physical quantity, and in this way, it accounts for how the physical quantity is located or arranged.

Moments are usually defined with respect to a fixed reference point; they deal with physical quantities located at some distance relative to that reference point. Magnetic moments are nearly synonymous[33] with the concept of a magnetic dipole, and you may hear these words used interchangeably.

A particle with a magnetic dipole moment is often referred to as a magnetic dipole. (A magnetic dipole may be thought of as a tiny bar magnet.) To date, singular magnetic charges that would be called magnetic monopoles have not yet

---

[33] We have a section on magnetic dipole moment in MRI because people have described the MRI physics through discussion of magnetic dipoles and magnetic moments. There may be subtle differences between concepts in the two nomenclatures, but effectively in the clinic and in translational research you may hear these terms nearly being synonymously applied to describe the sample and why that sample may have precession.

been found to exist in Nature [31]. Thus, a simple model that describes an infinitesimal loop coil has been proposed to be a reasonable approximation. The model magnetic dipole has a magnetic field as such a magnet and reacts to the influences of external magnetic fields. When placed in an external magnetic field, a magnetic dipole can be subjected to a torque that tends to align it with the field; if the external field is not uniform, the dipole also can be subjected to a force [39].

1.A.3.1 Magnetic moment
In MRI, we are using the magnetic moment, which is an advanced topic. We will provide some references in this appendix 1.A.3.1 for those who seek to have a more physics-based approach to understanding MRI (figure 1.55). As a 'simplified' model, you can think of moments as infinitesimal magnets combined in space (see magnet domains in appendix 1.A.3.2).

Magnetic moments are a fundamental concept in physics. Specifically, the term magnetic moment normally refers to a system's magnetic dipole moment, the component of the magnetic moment that can be represented by an equivalent magnetic dipole: a magnetic north and south pole separated by a very small distance.[34]

1.A.3.2 Learning about permanent magnets and domains
It is instructive to provide a little more information on magnetic domains, which are a major part of permanent magnets. The kind of magnets we use in MRI are typically of the electromagnetic type that we described in section 1.5.3, i.e., loops of coil that follow Faraday's law ($\Delta M \rightarrow E$) and Ampère's law ($\Delta E \rightarrow M$). However, there are some MRI systems that use permanent magnets, which we describe below.

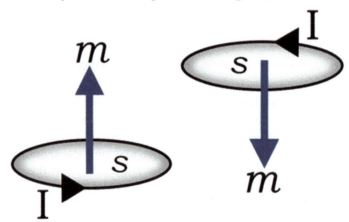

Figure 1.55. Magnetic dipoles can be represented as tiny magnets of microscopic to subatomic dimensions, and would have physics that result in equivalent behavior to an electric charge going around a loop. Rotating, positively-charged atomic nuclei all act as magnetic dipoles. Note: I = current, m = magnetic moment, s = surface area of the coil in this diagram.

---

[34] The magnetic dipole moment of an object can be expressed in terms of the torque that the object experiences in a given magnetic field. Recall that a dipole may be represented to be a vector. The same applied magnetic field (created by the main magnet field) creates larger torques on objects with larger magnetic moments. The strength (and direction) of this torque depends not only on the magnitude of the magnetic moment, but also on its orientation relative to the direction of the magnetic field, as will be discussed in greater detail in chapter 2.

Permanent magnets[35] typically are constructed by the magnetization of a piece of ferromagnetic material. Strange and wonderful is the world of magnetism. What happens in this world is that it divides into many small regions called magnetic domains. These domains are separated from each other by spatial boundaries called **domain walls**. The magnetization within each domain points in a uniform direction, but the magnetization of different domains may point in different directions.

A magnetic domain is highly relevant to permanent magnets, i.e., those magnets that have been exposed to a field and retain their strength without active energy. Note the picture of magnetic domains, as shown in figure 1.56. One fascinating component of these domains is that it seems you can keep on cutting them into smaller and smaller regions and they return this infinitesimal property, which is a region of dipoles that point in a direction; this region is known as a magnetic domain.

1.A.4 More on electricity and magnetism

Earlier in this chapter, we drew an analogy between the spinning proton and a more tangible spinning gyroscope. In fact, in class, we bring in physical models for the residents and technologists to see. This approach seems to help learners visualize the precession and helps them bridge the concepts into practice.

Once students have gained some confidence with the macro description of a gyroscope, the next learning opportunity applies the nature of imaging through a little loop of current generating a magnetic moment. Note that this simulation has intentional imperfections to illustrate the variation across a variety of theoretical spins. From visualizing an ensemble of rotating magnetic moments, we can also consider the appearance and construction of magnetic domains, as shown in figure 1.56 [36]. Ensembles of spins are magnetic moments that have angular

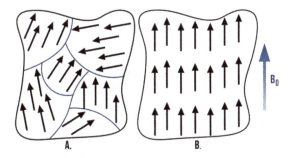

Figure 1.56. Magnetic domains are portions of magnetic materials (shown to the left). When exposed to a field, the dipoles can align, as shown to the right.

---

[35] Permanent magnets are similar to those that you may have worked or played a lot with when you were younger. They are energized by an initial magnetization and retain that magnetization for some time. Electromagnetics like loop coils with current that creates the magnetization through Maxwell equations (particularly Ampère's Law are a different form) are the primary form of magnet that is used in MRI to create the large magnetic fields we typically use in most MRI systems.

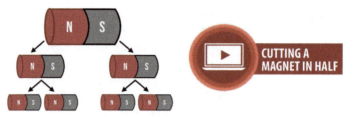

Figure 1.57. When a magnet is cut in half, the poles are not isolated. Cutting a magnet in half only produces more magnets. See video for a live experiment performed by the YouTube presenter [45].

Figure 1.58. A precessing gyroscope is shown, which may evoke memories from the reader's childhood. It definitely has been a part of mine and many others who received this 'serendipitous' gift when they were young [46]. Animation available at https://doi.org/10.1088/978-0-7503-1284-4. This 'File:Gyroscope precession.gif' image has been obtained by the author from the Wikimedia website, where it is stated to have been released into the public domain. It is included within this article on that basis.

momentum. It could be nice for some advanced readers to do a little physical analysis and comparison between the linear/straight-line components comparing effects, as shown in table 1.2.

To understand this, consider the relationship between force (mass × acceleration) to be the time derivative of linear momentum (mass × velocity)[36].

The analogy in rotation is the torque (which is an angular force) to have a time derivative equal to the angular momentum, figure 1.58. As seen in figure 1.59, the change in angular momentum of a gyroscope will be perpendicular to the torque by the right-hand rule. Thus, the gyroscope rotates around in a circle at a rate known as precession. Rather than look at equations, you can gather the sense of the motion of the mechanical gyroscope and translate those ideas to a spinning proton. That

---

[36] $F = ma$ and $p = mv$. Note p is the symbol for linear momentum. Also recall that acceleration is the time derivative of velocity and so the relationship between force and momentum is the derivative in time.

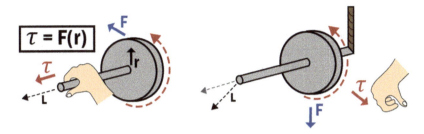

Figure 1.59. Shown in the figure is torque, with all the parts of the 'recipe' for torque in a gyroscope. This is important because soon we will make an analogy between the spinning top and a spinning proton (i.e., part of the hydrogen atom) [46].

Table 1.2. Comparison of formula between the linear frame and rotating frame for comparison.

	Linear/straight-line motion	Rotational motion
Distance	Δx	$\Delta \theta$
Velocity	v	ω
Acceleration	a	α
Inertia	M (mass)	I (rotational inertia)
Inertia x Acceleration	F (force)	τ (torque)
Momentum	$P = mv$	$L = I\omega$

Table 1.3. Elements required for precession in the E&M model with protons in MRI.

MRI/NMR elements needed for precession and spin of the 'proton'
Angular momentum intrinsic to atomic particles (the proton)
Magnetic field (main field), which must be very strong to get the number of spins to be converted to signal
Tilting magnetic moment. But what is going to be able to tip it? (see RF transmitter later in section 2.5 in chapter 2 on MRI hardware)

approach would be sufficient to glean the overarching concept of precession. This section attempted to help describe the thinking pattern that physicists and engineers understand when they think of precession.

A more detailed explanation of the physics of a gyroscope can be seen in [47].

Table 1.3 is a shorter table that focuses on the dipole moments and precession in MRI as a comparison with the mechanical derivation [47].

Some of this information can be difficult to understand, but serves as a PREVIEW for future concepts that will be discussed in this book. This is important for understanding the properties of matter that are interacting with E&M fields (important for T1 and T2 relaxation, which is a topic in chapter 3).

1.A.5 Populations of spins (parallel and anti-parallel spins)

Ever wonder how they discovered some of the fundamental portions of physics that are related to magnetic spin that are relevant for MRI? If so, please read on. Since we have discussed dipoles, then at this point it is good to introduce the reasons for the up and down states of spins. Understanding up and down spins (i.e., parallel and antiparallel spins) is further covered in the main field section of chapter 2 section 2.2.6. The following is a preview of that material.

As you recall, we described how protons have 'spin,' which is intrinsic. But in MRI, we are concerned with many protons. From a single spin, we will move to discussing what happens in an environment with multiple spins (a population of spins). Each spin aligns either parallel or antiparallel to the field. Each orientation has a different energy. The parallel spins have a lower energy state, while the antiparallel is in a higher state. If you add all the vectors of spins up, there will be a composite signal (isochromat) that will generate the net magnetic moment, which will be measurable by techniques like MRI. If you want to read more, please go to chapter 2, where we discuss this again in terms of the main magnet (figure 1.60).

Now the energy difference is a very slight difference between energy levels. This means that there is only a slight difference in the number of 'spins,' i.e., that will be between the lower and upper energy state. The ratio of differences in population can be on the order of tenths of a million. *This is why the number of molecules of water in the body is so vital in MRI. The high availability of protons (i.e., in water) is what permits there to be enough signal* (figure 1.61).

1.A.6 Introduction to the 'lattice' in MRI

At some point in your MRI journey, you'll likely hear the word or concept of 'lattice' being thrown around, particularly when you hear about spin-lattice relaxation, which

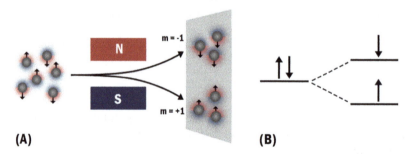

Figure 1.60. (A) For those curious about the historical experimental perspective, it is instructive to look at the original Stern–Gerlach experiment in 1921, in which a beam of silver atoms passed through a magnetic field [48]. The spot on the receiving plate was found to split into two, each having approximately the same size. The detected distributions were approximately half the intensity of a spot in the middle without the application of the magnetic field. This was the original experiment that illustrates a 'quantized' result. (B) Energy level representation of parallel and antiparallel spins. The 'minute' differences (on the order of parts per million) between the parallel and antiparallel spin distributions were determined later by another experiment, as shown in this figure [48].

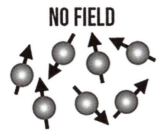 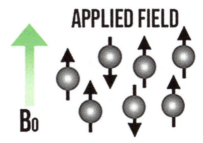

Figure 1.61. Application of a magnetic field B_0 aligns the spins parallel and antiparallel. Take-home message: in the presence of a strong and constant magnet field, alignment of spins creates two conditions: one in the direction of the main field (parallel) and the other in the opposite (antiparallel). The difference is small, but there are slightly more in the parallel direction.

is also known as T1-relaxation.[37] It is instructive to consider this physically relatable analogy with heat and exchange with the surrounding environment.

Natalie N. (premed & biomedical engineering student) thinks about the significance of the Stern-Gerlach experiment:
"I remember learning about this experiment in physics, and this experiment was significant because it proved some concepts that people had theorized for a long time. But why does the field matter? I understand that the particles split when they pass through the field and go into two locations as opposed to a single smear, but how does this apply to MRI? I'm having trouble seeing the big picture here."

Dr. Wu replies:
"Hi, Natalie. Yes, this may be hard to recognize, but it's the impetus for the parallel and anti-parallel spins that we talk about in chapter 2 (section 2.2.6). If it was a continuous smear of locations instead of the two that this experiment shows, then the response would be continuous. However, what was shown is that there are two different states into which spins are grouped by magnetization. Keep up the great thoughts and ideas. You are seeing lots of important concepts that we will revisit later and definitely 'up' and 'down' spins are an important concept in MRI to wrap your head around."

The lattice is as much a concept as it is a physical structure. It's easy to get lost in the details of how it 'works' and what it means, but we can describe the lattice as

---

[37] Synonyms for T1 relaxation include **longitudinal relaxation**, **thermal relaxation**, and **spin-lattice relaxation** (www.mriquestions.com) [49].

representing the environment or reservoir that surrounds the 'irradiated' object. Energy must therefore leave the spin system for T1 relaxation to occur.[38] This energy loss is unrecoverable and represents the transfer of heat; hence the origin of the T1 synonym *thermal relaxation.*' This energy must be transferred somewhere, and that 'somewhere' is into nearby nuclei, atoms, and molecules through collisions, rotations, or electromagnetic interactions.

T1-relaxation can be described through the energy exchange between spins and their external environment. This energy is dispersed quickly. Please recall the law of conservation of energy; energy needs to go somewhere so that the sum of all energies ends up being conserved. We will see the applied topic of relaxation as covered in chapter 3, but we will then proceed on to the next chapter, where we discuss five different hardware components of MRI.

Now that you have come to this point, have you ever thought about the conservation of energy and MRI? It turns out that MRI has a mechanism for exchange for spins with something called the lattice. As the energy returns to its equilibrium state, it has to send its energy somewhere.

When I describe this in class. we discuss the 'heat' analogy as described above, when you place a coffee cup with hot liquid on a table. We ask the students where they think the heat goes from the liquid as it cools down, as shown in figure 1.62.

Figure 1.62. Imagine a delicious cup of hot coffee (left) that you place on a table (reproduced from luckybusiness/stock.adobe.com). Note what happens to the heat energy if you leave the coffee on the table for some period of time. The heat gets exchanged with the surrounding environment (right) (reproduced from nikolya/stock.adobe.com). There is an analogy in MRI between the excited spins and where they can exchange their energy, which is called the 'lattice.'

---

[38] We will find in chapter 2 that we must excite the spins to get an image. Spins are the essence of signal in MRI. The excitation of these signals is part of the mechanism that produces the image contrast. But once they have been excited, where does all the energy end up going?

The answer is the 'environment.' The lattice is like that environment of spins with which the energy of spins can exchange energy after being excited. It's not a perfect analogy, but it provides the learner an opportunity to visualize a complex concept such as the lattice, which we describe below.

In physics, a lattice model describes a region as being a lattice as opposed to a continuous space. Lattice models originally occurred in the context of condensed matter physics, where the atoms of a crystal automatically form a lattice. A dipole lattice is the main physics structure that we will be considering, particularly with T1-relaxation. *First, recall the conservation of energy. We can describe the lattice as representing the environment or reservoir that surrounds the 'irradiated' object. (Note that this is a new concept here. The radiation is induced by RF energy, which in turn is transferred to the magnetic dipoles of the object, in our case, water.) However, the irradiated object will eventually decay, and this will then return the energy to its surroundings. The surroundings in our case is the 'lattice.' We will revisit this concept when we talk about T1-Relaxation in chapter* 3 (figure 1.63) [49].

Finally, this interlude provides a brief conceptual understanding that dipoles have directional influence, as well as influencing each other as an ensemble. Dipole interactions are the basis of relaxation parameters in MRI.

Summary:
1) We already looked at the impact of exposing the spins to a large magnet field (they align to states, parallel and antiparallel).
2) Consider what happens if you expose them to some energy that tips them away from their equilibrium positions.

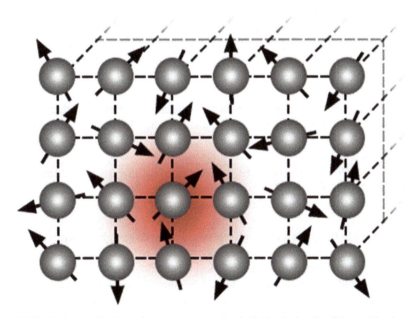

Figure 1.63. The lattice models the environment surrounding individual spins. It will be useful to have heard of this concept later when you study T1 relaxation (spin-lattice relaxation). Further, when spins attempt to return to the main field, the energy for that must go somewhere by conservation of energy.

3) However, when things are tipped away from equilibrium positions, they want to go back to their original states. Let us finally pause and review the coffee cup analogy. When the coffee cools down, where does the heat energy go? It must go somewhere by conservation of energy. Excited spins as they return to the equilibrium state will exchange energy with the aforementioned 'lattice.' This enables the conservation of energy of the entire system.

4) The lattice describes a 'theoretical space' of spins outside the sample. The sample naturally wants to return to its equilibrium position. It must exchange energy with something (by conservation of energy). Energy ends up being conserved if the spins in the sample that are returning to their equilibrium state in the field end up exchanging their energy with this 'lattice.'

References

[1] List of Nobel laureates in Physiology or Medicine https://en.wikipedia.org/w/index.php?title=List_of_Nobel_laureates_in_Physiology_or_Medicine…oldid=1065446407

[2] van Beek E J R *et al* 2019 Value of MRI in medicine: more than just another test? *J. Magn. Reson. Imag.* **49** e14–25

[3] Martinos T (n.d.) Center for Biomedical Imaging (Retrieved April 6, 2022) https://www.nmr.mgh.harvard.edu/training

[4] Friston K Statistical parametric mapping https://fil.ion.ucl.ac.uk/spm/

[5] Schild H H and Berlex Laboratories 1999 *MRI Made Easy (…well Almost)* (Wayne, NJ: Berlex Laboratories)

[6] Gaillard F and Baba Y 2021 Dural tail sign Radiopaedia.org (accessed 26 January 2022) https://doi.org/10.53347/rID-1247

[7] Khan Academy. Introduction to waves https://www.khanacademy.org/science/physics/mechanical-waves-and-sound/mechanical-waves/v/introduction-to-waves

[8] Khan Academy. Light: Electromagnetic waves, the electromagnetic spectrum and photons https://khanacademy.org/science/physics/light-waves/introduction-to-light-waves/a/light-and-the-electromagnetic-spectrum

[9] Encyclopaedia Britannica https://britannica.com/science/Maxwells-equations

[10] Magnetic Flux, Induction, and Faraday's Law. Boundless Physics (n.d.) (Retrieved April 6, 2022) https://courses.lumelearning.com/boundless-physics/chapter/magnetic-flux-induction-and-faradays-law/

[11] Callaghan P 2009 Introductory NMR & MRI: Video 01: Precession and Resonance https://youtu.be/7aRKAXD4dAg?t=122

[12] Boulder, U.O.C. Waves Intro (retrieved January 2022) https://phet.colorado.edu/en/simulation/waves-intro

[13] Stafanick G 2015 Frequency, cycle, wavelength, amplitude and phase https://blogs.arubanetworks.com/industries/frequency-cycle-wavelength-amplitude-and-phase/

[14] Aruba Blogs 2015 Frequency, cycle, wavelength, amplitude and phase https://blogs.aruba-networks.com/industries/frequency-cycle-wavelength-amplitude-and-phase/

[15] Inductiveload 2008 A diagram of the EM spectrum, showing the type, wavelength(with examples), frequency, the black body emission temperature. Temporary file for gauging response to an improved version of this file. Adapted from EM_Spectrum3-new.jpg, which is a NASA image https://commons.wikimedia.org/wiki/File:EM_Spectrum_Properties_it.svg

[16] Engineerguy 2014 (4/4) Operation: The details of setting up the Harmonic Analyzer https://www.youtube.com/watch?v=jfH-NbsmvD4

[17] 3Blue1Brown 2018 But what is the Fourier Transform? A visual introduction https://www.youtube.com/watch?v=spUNpyF58BY

[18] Gallagher T A, Nemeth A J and Hacein-Bey L 2008 An introduction to the Fourier transform: relationship to MRI *Am. J. Roentgenol.* **190** 1396–405

[19] Boulder, U.O.C. Fourier: Making waves https://phet.colorado.edu/sims/cheerpj/fourier/latest/fourier.html?simulation=fourier

[20] 2018 SmarterEveryDay What is a Fourier Series? (Explained by drawing circles)—Smarter Every Day 205 https://www.youtube.com/watch?v=ds0cmAV-Yek&t=52s

[21] MRI image formation - Objectives (n.d.) IMAIOS. Retrieved March 22, 2022 https://www.imaios.com/en/e-Courses/e-MRI/The-Physics-behind-it-all/Spatial-frequency-image-contrast-and-resolution

[22] User R (n.d). Water is Life. Root User. Retrieved March 22, 2022 https://www.watersnw.com.au/water-quality/education/exhibitions/water-for-life-exhibition/timeline2/water-is-life

[23] Boulder, U.O.C. Molecule Shapes: Basics https://phet.colorado.edu/sims/html/molecule-shapes-basics/latest/molecule-shapes-basics_en.html

[24] Water. Boundless Biology (Retrieved 26 January 2022) https://courses.lumenlearning.com/boundless-biology/chapter/water/

[25] Zaidan G and Morton C 2013 Why does ice float in water? https://youtube.com/watch?v=UukRgqzk-KE

[26] Khan Academy. Water, acids, and bases. Biology library. Science (retrieved 26 January 2022) https://khanacademy.org/science/biology/water-acids-and-bases/hydrogen-bonding-in-water/v/hydrogen-bonding-in-water?modal=1

[27] Milani G P, Bianchetti M G, Mazzoni M B M, Triulzi F, Mauri M C, Agostini C and Fossali E P 2014 Arterial hypertension and posterior reversible cerebral edema syndrome induced by risperidone *Pediatrics* **133** e771–4 https://pediatrics.aappublications.org/content/133/3/e771/tab-figures-data

[28] Why Is My Hair Staticky? How to Get Rid of Hair Static This Winter *Redken* (Retrieved April 6, 2022) https://www.redken.com/blog/haircare/what-causes-hair-static-and-7-ways-to-fight-it

[29] Rains H 2018 9 ways to stop getting static hair; from old wives' tales to must-have products *Good Housekeeping* http://www.goodhousekeeping.co.uk/fashion-beauty/hair-advice/a555691/how-to-stop-static-hair-products-tips-tricks/

[30] Rehm J and Biggs B 2021 Facts about four fundamental forces of nature Space.com https://www.space.com/four-fundamental-forces.html

[31] PBS. Movie: Einstein's Big Idea https://pbs.org/wgbh/nova/video/einsteins-big-idea/

[32] Faraday M and Royal Society (Great Britain) 1846 On new magnetic actions, and on the magnetic condition of all matter, continued c. 1 R. Soc. https://doi.org/10.5479/sil.389652.mq591311

[33] Khan Academy Electromagnetic induction (& Faraday's experiments) https://www.khanacademy.org/science/in-in-class10th-physics/in-in-magnetic-effects-of-electric-current/electromagnetic-induction/v/electromagnetic-induction-faradays-experiments

[34] Khan Academy What is Faraday's Law? https://www.khanacademy.org/science/physics/magnetic-forces-and-magnetic-fields/magnetic-flux-faradays-law/a/what-is-faradays-law

[35] K&J Magnetics 2015 K&J Magnetics–Electromagnet Experiment https://www.youtube.com/watch?v=9NniZSJL6Xs
[36] 7activestudio 2016 THE BAR MAGNET https://www.youtube.com/watch?v=DWQfL5IJTaQ
[37] Encyclopaedia Britannica Ampère's Law https://www.britannica.com/science/Amperes-law
[38] Brown R W, Cheng Y-C N, Haacke E M, Thompson M R and Venkatesan R 2014 *Magnetic Resonance Imaging: Physical Principles and Sequence Design* 2nd edn (New York: Wiley)
[39] Magnetic dipole | physics | Britannica (n.d.). Retrieved March 9, 2022 https://www.britannica.com/science/magnetic-dipole
[40] Khan Academy. Physical processes. MCAT. Test prep https://www.khanacademy.org/test-prep/mcat/physical-processes/magnetism-mcat/a/usingthe-right-hand-rule
[41] Westbrook C and Talbot J 2018 *MRI in Practice* 5th edn (New York: Blackwell)
[42] Kohama Y, Ishikawa H, Matsuo A, Kindo K, Shannon N and Hiroi Z 2019 Possible observation of quantum spin-nematic phase in a frustrated magnet *Proc. Natl Acad. Sci. USA* **116** 10686–90
[43] AstroAI Multimeter 2000 Counts Digital Multimeter with DC AC Voltmeter and Ohm Volt Amp Tester; Measures Voltage, Current, Resistance; Tests Live Wire, Continuity - Amazon.com. (n.d.). Retrieved February 16, 2022 https://www.amazon.com/AstroAI-Digital-Multimeter-Voltage-Tester/dp/B01ISAMUA6/ref=sr_1_5?dchild=1&keywords=multimeter&qid=1595721751&sr=8%E2%80%935
[44] Amazon.com: Uxcell DC 6V 500RPM Micro Speed Reduction Motor Mini Gear Box with 2 Terminals for RC Car Robot Model DIY Engine: Toys & Games. (n.d.). Retrieved February 16, 2022 https://www.amazon.com/uxcell-18000RPM-Speed-Micro-Motor/dp/B00TX2Z6OQ/ref=sr_1_60?dchild=1&keywords=Motor&qid=1595722026&sr=8%E2%80%9360
[45] QM Oxford 2019 Magnetism: Cutting a magnet in half https://youtu.be/q0d_SS-0Puk?t=162
[46] Lumen University Physics Volume 1. 11.4 Precession of a gyroscope https://courses.lumenlearning.com/suny-osuniversityphysics/chapter/11-3-precession-of-a-gyroscope/
[47] van Biezen M 2016 Physics - Mechanics: The Gyroscope (3 of 5) The Torque of a Spinning Gyroscope https://www.youtube.com/watch?v=qS_dcNqs3d4
[48] Gerlach W and Stern O 1922 Der experimentelle Nachweis der Richtungsquantelung im Magnetfeld *Zeit. Phys.* **9** 349–52
[49] T1 relaxation (n.d.) Questions and Answers in MRI. Retrieved April 6, 2022 http://mriquestions.com/what-is-t1.html
[50] Christian W and Forinash K 2018 Adding two linear waves (superposition) https://compadre.org/osp/EJSS/4030/138.htm

Chapter 2

Hardware: five important components and beyond

(Main field, shim, gradient, RF transmitter, and RF receiver)

2.1 Introduction

We will discuss five main hardware components in MRI. These will be divided into:

1) **Main magnetic field**—permits precession and establishes the signal levels;
2) **Shim**—reduces unintended spatial gradients in the main field;
3) **Magnetic gradient**—establishes spatial localization;
4) **RF transmitter**—tips the spins, initiating precession;
5) **RF receiver**—listens to the signal.

Throughout the rest of chapter two, we personify the hardware as having 'jobs.'[1] It is easiest to organize each component by focusing on their unique responsibilities (figure 2.1). The hardware chapter builds upon the five core concepts introduced in the first chapter. We shall indicate where key principles rely on any of the above main hardware components so that you can build upon them.

---

[1] We use the 'personification' of the hardware as having a job to help the reader think about the responsibilities and application of each piece of equipment.

Figure 2.1. When you first start learning MRI hardware, it might often feel like looking at a coil within coil, yet another coil, and so forth. Much like the matryoshka dolls pictured above, this icon will be used as a cue for when a concept concerns MRI hardware. These images have been obtained and adapted by the author from the Pixabay website where they were made available under the Pixabay License. It is included within this article on that basis.

Amy B. (psychology graduate student), reflecting on the principles of changing electric and magnetic fields, aka Faraday/Ampere Laws, said: "To prepare for this section, I reminded myself of the information in chapter 1 about Faraday's Law and Ampère's Law. Electricity and magnetism are related, such that if you change the electric field, you generate a magnetic field, and if you change the magnetic field, you generate an electric field. In a distant memory, I recalled seeing my professor bring out some coils and put some current in it and it produced a magnetic field. So, if you have electricity flowing through coils, this can generate a magnetic field, and vice versa. This is probably important to understanding how the MRI equipment works."

Dr. Wu responded:
"In chapter 1, you discussed both Faraday and Ampère's law in regards to changing electric fields making a magnetic field and changing magnetic field making an electric field. These concepts will come up repeatedly in this chapter, as the magnet hardware itself is composed of many coils and loops and involves the transduction of signal between different magnetic and or electrical sources to convert them to other measurable factors."

A basic understanding of the hardware will help with later perspectives, including relaxation, signal-to-noise ratio, pulse sequences, and artifacts. The emphasis on MRI hardware within radiological board examinations varies in content between years. As a start, we encourage you to memorize the roles of each hardware from table 2.1 below. This will provide you with a better sense of what is going on in a complex instrument to help your patients, as well as have a better intuition of what you are 'driving' and what creates the images that healthcare workers will look at on a day-to-day basis. Some of you will have access to technical expertise, such as an MR Physicist. If so, it can be advantageous to work with them to develop your program not only for quality assurance (QA), but also for MR protocols, technological applications, and MR safety programs. Regardless, the understanding of these concepts can help you to be more informed about the use and reading of MRI.

2.1.1 Ten reasons why it may be good to start with a perspective on MRI hardware:
1. It's a concrete (tangible) and relatable item within MRI technology (less abstraction).
2. It's the unit that functions as the 'camera' from which we obtain images.
3. It's what technologists work with every day.
4. If you are a medical physicist, it can be important for MRQA and working to help resolve service issues.
5. For biomedical engineers, understanding hardware is critical so that you can develop or provide appropriate assistance to the users of the system.
6. For physicians, it permits you better communication with technologists and setting up protocols to get the correct images needed.
7. For researchers (psychologists, rehabilitation, neuroscience, etc), it can be an exciting area for investigating function, metabolism, pathophysiology, and even assessing the mind.
8. While scanners are less what physicians may interact with on an everyday basis, it is important to understand hardware, as physicians hold responsibility for MRI safety and improving protocols and scanner efficiency.
9. It can be a gateway to understanding artifacts and can be important when working with service (especially if there is no MR physicist or scientist available at your institution).
10. When purchasing scanners, it is important to understand the technology.

Nicholas P. (third-year medical student) reflects:
"I think the most important reason to understand MRI, really, is responsibility of physicians. Scanners are not what physicians may interact with on an everyday basis. However, it is important to understand hardware, as physicians hold responsibility for MRI safety and improving protocols and scanner efficiency. I am ready to learn the five concepts! Seems like it will help my understanding of MRI, which will be beneficial in my future career."

2.1.2 Yield bubble

For the IMPATIENT reader (for example, those whose boards are fast approaching) who wants to get 'cracking' on the basic building blocks of MRI and beyond, which are the topics covered in the next chapters, consider first looking over the concepts in table 2.1 below. This table contains a summary of concepts that will connect the dots for you in your future career. Please consider scanning the summaries for each section, and then proceed to the next chapter that will begin with discussions on resolution, relaxation (T1, T2, T2*), and pulse sequence, all of which are highly relevant to a clinical MRI practice. You can always look back at these materials if you find that you desire a more conceptual understanding of hardware for reasons listed in the above table.

As a note, we usually cover the first two chapters in two-to-three lectures with first-year residents, and we also discuss tradeoffs and artifacts. For second-year residents, we start with the third chapter after answering any basic questions they may have prior to the start of lectures. Other disciplines (technologists, medical physics, medical students, biomedical researchers interested in MRI, and other subspecialties) will have more time to spend on the topic and have more responsibility for the concepts in the first two chapters. So, we encourage those other professions to start with the first two chapters.

Table 2.1. Roles/jobs for different hardware in MRI. The author created a simple summary that can be used by the 'busy' professional, and those that want to continue forward to chapter 3, but may come back to look at these items at a later point.

1. *Main Field:* aligns spins and sets the resonance frequency.
2. *Shim:* reduces inhomogeneous spatial fields and reduces unintended spatial gradient.
3. *Gradient:* provides spatial encoding.
4. *Radiofrequency transmitter:* tips spins by brief magnetic pulse.
5. *Radiofrequency receiver:* listens via an RF coil.

2.1.3 Brief overview of the anatomy of the scanner

It is instructive to continue to visualize the anatomy of the magnet and its parts. At this point, we also have layered in the relative positions for some of the 'shim' coils. This approach is definitely useful for radiologic technologists, medical physics, and research engineers. For physicians, it can only benefit you to know about the machine that is going to enable you to better select equipment that you buy, as well as to manage safety for which you are responsible in the clinic.

Matryoshka dolls, or Russian dolls, are a set of wooden dolls of decreasing size that are placed one inside another. The name matryoshka literally means 'little matron.'

MRI coils (main magnet, gradients, RF transmitter, RF receiver coils) are coils within coils within coils, much like the matryoshka doll.

2.2 Main magnet
Learning objectives for main magnet

 1) State the different forms of magnets in MRI.

 2) State where the location of main field coil is within the scanner.

 3) Understand the role of the solenoid in generating the main field (magnet).

 4) Understand how the main field serves to align spins.

 5) Understand how the main magnet field permits precession.

 6) Understand how to calculate resonance frequency for water based on magnetic field strength.

2.2.1 Different MRI magnet types

Let's start by recognizing there are different MRI magnet types (three described below):
- Permanent magnets—capable of sustaining fields up to about 0.3 Tesla (see section 1.A.3.2);
- Resistive magnets– capable of fields up to about 0.6 Tesla.
- Superconducting magnet.

If you are curious to read more, consider looking at MRI-Q.com.

The most common type of magnet in clinical MRI is the **superconducting magnet**. We will focus most on this type of magnet. The superconducting MRI coil is an electromagnet (see section 1.5.3) made from coils of specialized superconducting wire[2]. Typically the electromagnets are 'solenoidal' in structure, which permits a magnetic field to be in the middle of the magnet.

Remember a solenoid is a coil of wire, usually in cylindrical form, that when carrying a current acts like a magnet. You already have two environmental factors to induce precession—the main magnetic field and the intrinsic spin of the proton. The RF transmitter adds the final piece by creating a magnetic pulse to tip the protons (located in the position in figure 2.19). The magnetic field from the RF transmitter is called B1. It will typically operate at extremely low temperatures (just several degrees Kelvin, which is approximately –273 °C). In the superconducting state, there is minimal electric resistance, which can result in a stronger magnetic field. The scale of the magnetic fields that we use today is on the order of 1 Tesla. Many are 3 Tesla, but most common is 1.5 Tesla. To date, there are up to 7 Tesla magnets available for clinical use, and as of 2022 there is some interest in exploring 7 Tesla utility for focal cortical dysplasia, which includes epilepsy and multiple sclerosis. In terms of units and scale, recall that 1 Tesla is approximately at 20,000 times the Earth's magnetic field[3]. (See section 1.A.2 where we discussed scales of magnetic units.) That's a lot of stored energy!

The main magnetic field, also known as the static field, is commonly referred to as the B_0 field in MRI scanners.

Let us next perform a Deductive Learning Scene (DLS) by looking at a photograph of a 1.5 Tesla scanner. Examine figure 2.2 and imagine the location of main field coils. Visualize the different types of coils; remember that we referred to this as being analogous to the matryoshka dolls[4]. Please write down your thoughts and try to connect with the five core concepts throughout the rest of the chapter.

---

[2] See section on electromagnetics in section 1.5.3 in chapter 1 that has a brief introduction to electromagnets. Note, there are also permanent magnets that are used in the shim process, but not necessarily the main magnet, which is typically a superconducting electromagnet. Note that there are permanent magnets that you will encounter having this form in other applications (not necessarily MRI). We briefly discuss magnet domains, which are the fundamental part of the physics of how permanent magnets work, which was discussed in section 1.A.3 of the chapter 1 appendix.
[3] See section 1.A.2 for more details about the scales of magnetic fields in the appendix of chapter 1.
[4] This matryoska dolls image has been obtained by the author from the Pixabay website where it was made available under the Pixabay License. It is included within this article on that basis.

Figure 2.2. Deductive Learning Scene (DLS) of MRI scanner.

2.2.2 Suggestions on DLS and how to think about the main magnet field

As you are going through this chapter, and you are thinking about for the above Deductive Learning Sketch in figure 2.2).

1. The field strength is 1.5 T, therefore it is a superconducting magnet.
2. The main magnet will have a lot of coils. More loops make a stronger field.
3. Current is running through the coils to create magnetic field. Recognize that $\Delta E \rightarrow M$.
4. A magnetic field develops within the scanner due to a solenoidal organization of coils.
5. You may observe the table that will slide the patient in.
6. There are multiple types of coils present with different 'jobs.' We will address the gradient coils, shims, and RF coils in detail later.

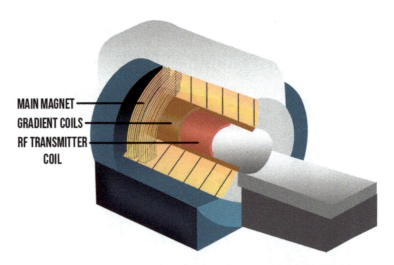

Figure 2.3. This figure shows the general location of various elements of the coils: (1) Main magnet with large number of loops on outside; (2) gradient coils which provide the spatial localization in MRI; (3) RF transmitter coil to excite the 'protons' of the hydrogen contained in 'water' of the patient. *Not pictured are the Shims and RF receive coil, which will be covered later in the chapter.*

2.2.3 Location of the main field

The main field, or commonly referred to as the B_0 field, is constructed on the principle of a solenoid. Looking at figure 2.3 above, we get a peek at the many repeating coils. Together these coils make up the main magnet. The coils create a homogenous, uniform magnetic field, as can be observed in figure 2.4.

Figure 2.4. Approximate location of main magnet.

2.2.4 Role of the solenoid

A solenoid employs a current which changes direction as it turns through the winding. This 'change' produces magnetic field lines. Solenoids are important as they create magnetic fields that are used as electromagnets. Solenoids can be specifically designed to produce a uniform magnetic field in a volume of space, which is important for MRI. A solenoid has a changing electrical field (in terms of geometrically turning spatially) that produces magnetic field lines within its structure. We refer to this as an electromagnet (discussed briefly in section 1.5.3). The solenoid is formed from multiple coils of wire with an electrical current running within the wire, as seen in figure 2.5. A uniform magnetic field is created within the center of the coils. The magnetic field within the solenoid can be strengthened by increasing the amount of coils and/or increasing the current, as depicted in figure 2.6. Hence, 3 T magnets are bulkier than 1.5 Tesla magnets, having a larger number of loops that are capable of generating a larger magnetic field.

If the coils of the solenoid are not tightly packed, the main field will bend. **It is crucial the main magnetic field of an MRI be flat and uniform.** This is because *unintended* gradients will cause signal loss and increase difficulty in 'mapping' intended gradients, resulting in distortion of the image[5] [1]. Shimming methods can be used to help flatten the field (i.e., active and passive shimming). See figure 2.7 for more details.

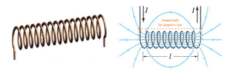

Figure 2.5. (Left) a three-dimensional representation of a solenoid (this Solenoid-1 image has been obtained by the author from the Wikimedia website, where it is stated to have been released into the public domain. It is included within this article on that basis); and (right) a 2D picture of a solenoid with injected current and magnetic field lines (this 'File:Solenoid and Ampere Law - 2.png' image has been obtained by the author from the Wikimedia website where it was made available by User:Goodphy under a CC BY-SA 4.0 licence. It is included within this article on that basis. It is attributed to User:Goodphy).

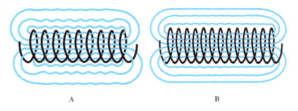

Figure 2.6. With an increase in number and pitch of these turns in the solenoid, the field becomes more 'flat'. Notice how solenoid coils are used to create a field that is mostly uniform inside the bore of the magnet in B, while with a lower pitch and decrease in the number of coils there is the possibility of more [2].

---

[5] A side note for neuroscience and brain researchers. One key concern for these researchers revolves around the type of sequences used in functional MRI that are susceptible to distortion. *Unintended* gradients can cause signal loss and spatial position distortions, which can alter interpretation. For example, optimization of magnetic field homogeneity in the human is reviewed by Koch *et al* [1] for those interested in some sophisticated technical details that your team may consider investigating to optimize research results.

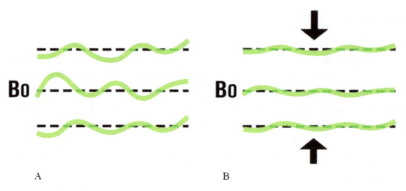

Figure 2.7. Illustration of the goal of trying to straighten magnet fields ('flattening the field'). Note that due to imperfections there can be oscillations in the field. The goal of 'shimming' is to reduce the fluctuations to minimal levels. The goal is to achieve straight fields, or a uniform gradient (the idealized dotted black lines) throughout the imaging volume. Without proper shimming, you may have additional unintended fluctuations in the field, as shown in green. Shimming methods are used to minimize and reduce these fluctuations and improve image quality. Black arrows in figure B are meant to indicate the direction in which fields would be driven to accomplish a 'flatter' field.

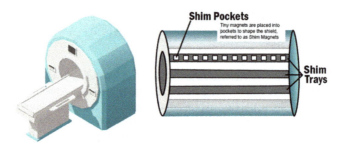

Figure 2.8. One form of magnetic field shimming involves the use of small permanent magnets placed into pockets. These pockets are placed in shim trays which are slid into the bore to aid in creating a more homogeneous field over a volume. Several sized shim volumes are specified to be optimized for homogeneity. For more information on passive shimming see MRIQuestions.com. Left image from Taras Livyy/stock.adobe.com.

You have already learned that spatially varying fields are called gradients. You have learned that implementing 'gradients', can encode spatial position. Recognize now that if you have unintended gradients, then you will generate an encoding of the location of spins.

These altered local fields will cause distortion in images (figure 2.9).

2.2.5 Summary to this point

- Note that there can be a field created in the middle of the electromagnet (where you can put the patient).
- You can design the lines to be straight, especially in the center of the cylinder.

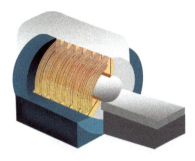 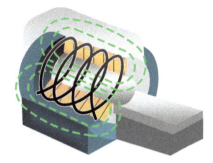

Figure 2.9. Note in the left image the position of coils in the MRI system. Note the solenoid superimposed onto the magnet to help visualize the role of the solenoid in the field.

- Finally, remember what the word shimming means (straightening), and that additional methods can be used to flatten the field. The methods include applying passive shim, target area, and active shimming, all of which can improve the performance of the magnet.

Carlie P. (chemistry/psychology premedical student and dancer) reflected on solenoids:
"Solenoids are interesting because they enable you to get flat magnetic fields through changing electric fields. I think that this change in electric fields comes from the directional shift of the current within the wire coils. If you have more coils, you can generate more electric field changes and therefore a higher and more homogenous magnetic field. Engineers who develop MRI magnets must have a very difficult job in making sure to place enough coils in the correct places so that the magnetic field does not bend."

2.2.6 Alignment of spins

We recall protons have polarity and spin, allowing them to act like tiny magnets. Naturally, we have no net magnetization because our protons are aligned in many different directions. During an MRI, the 'job' of the main magnetic field (B_0) is to force the protons to align along the same axis, creating a net magnetization—just like a compass aligns with the Earth's magnetic field.

The spins can align in two orientations along the axis. The majority of the spins will align in the same direction, parallel, of the main field (up-spin). However, some will align against, anti-parallel, to the main field (down-spin)[6]. Stronger magnetic fields create a larger difference in the up versus down spin ratio and thus produce more observable differences, resulting in a stronger signal, as can be seen in figure 2.10.

---

[6] We touched on the topic of alignment of spins in appendix 1.A.5 in chapter 1, where we described the Stern-Gerlach experiment, which provided the breakthrough evidence on the existence of two states of parallel and anti-parallel spins.

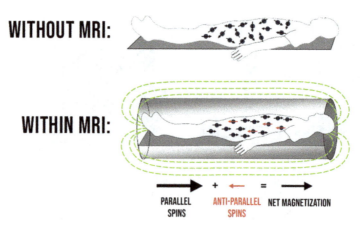

Figure 2.10. Without the presence of the external main field (B_0 field), the dipoles orient in many different directions, resulting in no net measurable signal. When the main field is applied, the spins align along the main field, creating a net magnetization. Note: slightly more spins align parallel as compared to anti-parallel.

2.2.7 Fun-fact bubble: Analogy for the two states in MRI for 'parallel' and 'antiparallel' spins

Here is a 'fun fact' to understand alignment and energy differences:

You return home from a long shift at the hospital. You can choose to rest in a comfy chair (a lower energy state) or study at the table (a higher energy state), which is depicted in figure 2.11. Would more residents choose to rest or study? This would depend on your energy level. Some will choose the high energy state of studying, but most will choose the lower energy state of rest.

Figure 2.11. Higher and lower energy states analogy depicted through chairs. These energy states will relate to the number/population available in MRI for the 'parallel/down' and 'antiparallel/up' states of spins.

The choice of which chair you sit in would depend on the 'energy' difference. If there was very little difference and the desirability of the two chairs was the same, then the number of people choosing each chair would be split nearly down the middle.

> However, if there was even a tiny difference in energy, you would have enough people selecting a 'sitting' position or chair to the point that you would start to see an observable difference in population/numbers between chairs. The chair analogy provides a metaphor that is meant to describe how to help the reader conceptualize 'available' spins[7] between two states. Further, this model illustrates the distribution of spins in a main field where two choices are available. If you read chapter 1, appendix 1.A.4 you may recall the Stern–Gerlach experiment, which illustrated experimental evidence of the two-state model for spins in the presence of a magnet field. In MRI, the population of 'available' spins from hydrogen is large in number. However, the energy differences are extremely small between the parallel and antiparallel states. This analogy/metaphor was meant to help the readership gain some intuitions about why hydrogen is a good target for MRI (if water is the target, because of a large number of 'available' spins). This number of available spins is also important because the fractional difference created by the two states (parallel and antiparallel spins) is a tiny number and must be overcome to generate an adequate signal in the image.

Protons in 'up-spin' are in a low energy state, protons in 'down spin' are in a higher energy state. The energy difference is very small and so results in differences in the population of spins up and down. This is a very small fraction of the number of spins exposed to the field. The fractional difference is on the scale of tens of parts per million (~1e-7 or tenths of parts per million [ppm])[8]. The difference in the spins pointing up and those pointing down will depend on the total number of spins that react to the static field (in MRI parlance, the static field is referred to as B_0).

The main concept to consider is that there is a slight percentage more (1e-7) in the parallel versus the anti-parallel. It is the greater number of available spins as the number of spins shifted to the parallel and antiparallel direction that creates the source for the signal. It will be the difference from those spins that will generate the final signal available to produce an image.

However, in the case of hydrogen, we have so many available spins. Recall that we are mostly comprised of water. In a typical voxel[9], there is an abundance of available spins: $N = 10^{15}$ spins. We refer to available spins as the number of spins that respond to a static magnetic field and align with the field. Note that with a large number of available spins, the small fraction becomes a number that is a usable value[10]. Finally, as a point of understanding more MRI terms/vocabulary, when we can add vectors and the total addition of spins, both up and down together, this will be called an **isochromat**.

---

[7] Some visualization of **nuclei (protons)**, which have the magnetic properties, called **nuclear spin**. The authors at Imaios described the behavior as like tiny rotating magnets, represented by vectors [3].
[8] The difference between the up and down states is relatively small. Thus, you need a lot of available protons to generate a usable signal, because the energy difference between the states is very tiny.
[9] A voxel is a three-dimensional entity that represents part of the image.
[10] To better understand this, multiply the number of available spins by the fractional difference between the probability of being in a state and examine the difference (parallel versus antiparallel spins).

2.2.8 Scaffolding bubble

It is a little instructive to do a few calculations and review some chemistry concepts to provide a sense of scale for MRI hydrogen molecules and populations participating in the available signal in MRI. The signal that is created by a system that contains the high and low energy states will be expressed as the energy difference between up and down spins, as can be visualized in figure 2.12. If, for example, we plug in for $N = 10^{15}$, for an estimate of the number of spins in a 1 c.c. of water, we get $N*(1/10\,000\,000) = \sim 10^{15}/10^{7}$, which has sufficient spins contributing to the signal. Note, this is the sum of the two populations of dipoles, in the case of MRI as seen on the left.

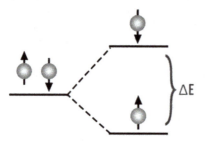 

Figure 2.12. Protons shift between higher energy states (down spin) and lower energy states (up spin). There is a stronger impetus to be in the lower energy state, but as this state is filled, protons are forced to occupy the higher energy state.

Even though 10^8 sounds like a large number, you still would need this number of differences in spins to get enough signal for diagnostic purposes.[11]

Gail H. (post doc neuroscientist) reflected on parallel and antiparallel spins: "This section helped me understand how there is a signal in a magnetic field. The proton spins are aligned along the direction of the magnetic field, but some face 'up' and some face 'down.' The reason the signals don't cancel out is because the distribution between 'up' or 'down' isn't equal, so there is a net signal. The population distribution depends on if the spin is 'up' or 'down,' and the protons prefer to be in the lower energy state. This reminds me of chemistry class when we talked about electrons having energy states. This is interesting, as it begins to explain the reasons for 'up' or 'down' spins in MRI. I guess that has something to do with the available SNR and will read on to find out more."

Summary

Without the main field, the spins (magnetic dipoles) would orient in random directions. However, this pure randomness of orientation would not generate a signal. This is one of the reasons why we place a human body into a very strong

---

[11] You need a lot of available protons to generate a usable signal because the energy differences are tiny, as we mentioned previously in this section.

magnetic field (i.e., like that produced by the MRI scanner). Just like a compass needle aligns to the Earth's magnetic field, when these randomly spinning hydrogen protons are placed in an MRI scanner, their axes realign them with the scanner's stronger magnetic field. We call the scanner's magnetic field the B_0 field. Also note that B1 refers to the magnetic field pulse produced by the RF transmitter that tips these spins; we cover the RF transmitters' hardware in more detail in chapter 5.

2.2.9 Permit precession

As a brief review on the topic, to understand precession, keep in mind as you think about the role of each hardware piece in MRI that there should be a magnetic moment with spin sitting in a static field. There will not be precession without tipping the spins, similar to how we need tipping for a gyroscope to yield observable precess. For a mechanical gyroscope to have precession, we need gravity. Here, the main field functions like gravity, but in an EM sense. We refer to 'permit' precession as an environment created that will make conditions possible for precession; however, permitting the precession still requires additional energy to perturb or start the process of precession[12].

2.2.10 Precession revisited and Larmor frequency

The next 'job' of the main field is that it sets the resonance frequency[13]. To compute this, we will use the famous Larmor formula: **Field strength dictates operating frequency**. The connection between the frequency and field follows below (figure 2.13).

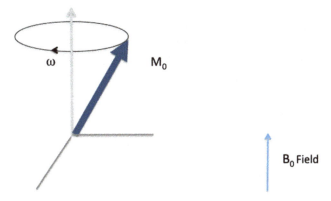

Figure 2.13. Recall from chapter 1 the concept of precession, and look at the scaffolding bubble on populations. In chapter 1, we understood how the main field permits precession. It is one of three key elements: (1) the intrinsic spin of the proton; (2) the presence of the spins; and (3) the ability to tip the spins. The first two we have already covered, while the third item in that list will be described in the RF transmitter section.

---

[12] In a gyroscope, to have precession (mechanical) you need gravity. In the MRI model, remember we need the main field to act as one of the ingredients needed to have precession (the other ingredient is energy to 'tip' the spins) which is a different piece of hardware.

[13] Resonance is a phenomenon characterized by an oscillating response to external input. In MRI, the primary source of external input is produced by the RF Transmitter (B1). The energy of the B1 is sent to the rotating sample. Transmitting on resonance that is at Larmor's frequency will maximize the energy transferred by the transmitter, while off-resonance (perhaps even slightly off that Larmor frequency) would be a less efficient transfer (Precession on mriquestions [4]).

$$\omega = \gamma B \tag{2.1}$$

where ω = the Lamor frequency, γ = constant which is known as the gyromagnetic radio, and B is the field strength.

The Larmor equation is provided above. It is next instructive to compute the resonance frequency for several strengths of magnets. At 1.0 T, we operate at ~42.6 MHz, 1.5 T at 63.8 MHz, and at 3 T this is 127.6 MHz. Note that the gyromagnetic ratio for water is = 42.577 MHz/T.

When you think about the field, think about a linear relationship related through a constant. The Larmor equation provides the link between magnetic field and frequency, which is a fundamental piece to understanding MRI.

Note that the main magnet is not the only thing that you need to have precession. In the gyroscope model, something will have to tip the gyroscope. So, that observable precession will occur where we have the RF transmitter to create that tipping. This is why we refer to one of the main magnet's jobs as to 'permit' precession, as it creates the environment for which the RF transmitter can tip. For those who are curious, the other element that causes the tipping is the RF transmitter, which we will talk about later in this chapter in section 2.5.

2.2.11 Summary of the main field

A simplified description that may be enjoyed by a broad set of readers is contained in the open access article by K Broadhouse titled 'The physics of MRI and how we use it to reveal the mysteries of the mind' on kidsfrontiersin.org [5]. Think of a compass needle in the Earth's field. The compass itself does not physically move, but the needle spins to align itself. Similarly, the hydrogen protons do not physically move in your body when you enter an MRI scanner; their axes just align along the direction of the B_0 field. Some will align 'up' (parallel) and some will align 'down' (anti-parallel), while still spinning around on their own axes. Due to the laws of quantum physics, which we will not go into too much here, there are always just slightly more 'up' protons than 'down.' If you now think about the total magnet field generated from all of our hydrogen protons, these tiny magnets almost cancel each other out, to leave only the magnetic field from the small proportion of extra 'up' protons, and it is this small magnetic field that we can measure using MRI. To learn more and review this concept, readers may like to review/look at appendix 1.A.3, where the original observation of this 'split' was first discovered experimentally.

Note that the precessional frequency (when we talk about **precessional frequency**, we also refer to that frequency as being the **Larmor frequency**, after the formula we use to calculate the rate of precession [6]) is given in equation 2.1, which depended on the strength of the magnetic field and gyromagnetic ratio where the latter is dependent on the chemical species we are operating at (water). The stronger the magnetic field, the faster the protons spin. *These two ideas of axis realignment and precessional frequency are important when we use MRI to measure the signal from these hydrogen molecules.*

2.3 Shim coils

2.3.1 The goals of shimming (reduce unintended gradients)

As we continue looking at different parts of the MRI scanner hardware, we are systematically proceeding from the outside of the hardware cylinders to the inside, as shown in figure 2.14 below. The next piece of hardware is the active shim coils.

Shimming is a tool that assists in flattening the main field by reducing unintended gradients. Recall that when we see a term that has to do with spatial gradient, we should try to see if it has to do with spatial localization.

However, even a slightly off field will cause unintended signals in the image. So, we do not want any gradients other than the one that we specified for spatial localization. This is because those 'extra' gradients will mismap the locations, causing error, as illustrated in figure 2.7.

Figure 2.14. Shims are included in our diagram. Shims are usually placed around the main magnet in order to make the field more homogeneous. There are several concepts of shimming, such as active shim (uses electromagnetics) and passive shim (bar magnets).

Language bubble

SHIM
A: 'a tool, such as a thin strip of material, used to align/ make things straight.' [7] (figure 2.15)
B: to fill out or level with use of a shim (figure 2.11)
(figure 2.12)

Figure 2.15. Shim, explained as the adjustments we make to furniture like a wobbly table in a favorite mom and pop restaurant that is present in many small towns and cities across the world. This 'File: Downtown-guthrie-oklahoma.jpg' image (right) has been obtained by the author from the Wikimedia website where it was made available User:Kkinderunder a CC BY-SA 3.0 licence. It is included within this article on that basis. It is attributed to User:Kkinder.

2.3.2. Scaffolding bubble

Question: Why is it difficult to 'straighten' the magnet field? And what do poor shims look like?

The field is large, being 20,000 times the magnetic field of earth. It is difficult to achieve precision. Note that, because any variations of straight lines will actually produce variations in the level of magnet field, this in turn will cause mis-mapping within image location, as shown in figure 2.16. To understand more, see MRI Questions on the topic [8][14].

Figure 2.16. If the system is not shimmed properly, then distortions will appear in the image, such as in the top image where the lines should be rectilinear, while in the bottom figure mismapping of the head shape will occur when field lines are not made to be more homogeneous.

---

[14] In section 2.4 we will discuss further the concept of the spatial gradient. Gradient hardware is used in MRI which creates a changing magnetic field per space. In section 2.4, the role of the gradient is described as producing spatial encoding. In the case of poor shim quality, the result will produce unintended gradients (i.e., unintended magnet field that varies over space. This effect will produce unintended spatial positioning in an image and thus lead to signal distortions. The role of the shims is to minimize such distortions in the image so that the hardware gradients are producing the best possible image positioning.

Summary
1. Note that there can be a field created in the middle of the electromagnet (where you can put the patient).
2. You can design the lines to be straight, especially in the center of the cylinder.
3. Finally, remember what the word shimming means (straightening) and that the additional things that can be used to flatten the field are described in the appendices. The topics there include passive shim, target area, and active shimming, all of which can improve the performance of the magnet.

Whitney (MR Technologist) explains:
"Radiology techs do everything to try to get the best, and straightest, magnetic field. First, they make sure the coil is centered on the correct anatomy and centered in the bore, or opening of the machine. Next, they remove any metal from the patient and reduce any inhomogeneity through shimming. They then run the sequence which is most appropriate."

2.4 MRI gradient coils

Key learning objective on the gradient:

Understand that the KEY ROLE of the 'Gradient Coils' is to provide **spatial localization**. Also note that there is specific MRI hardware that permits gradients to be created in any direction (x, y, z). The gradients can also be placed in any vectorial combination so that the orientation of the acquisition can be in any plane. The reader should identify gradients in spatial localization. Additional usages that gradients can have are the vectorial directions they can be oriented in, and the pattern in which the gradient lines will be seen to traverse k-space. We will discuss in chapter 4 the pulse sequence diagram in more detail. The specification of 'playout' of various gradients in the pulse sequence diagram along with signal echos will create (as discussed in chapter 3) the desired T1, T2, or other 'weightings'. Application of the spatial gradient was the prime concept from which Paul Lauterbur (of Nobel Prize fame) built his ideas, transforming the concept and possibility that now provides us with a methodology to obtain medical images with magnetic resonance that were previously not practical, as depicted in figure 2.17.

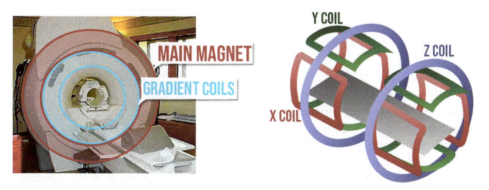

Figure 2.17. Left, the position of the gradients. Right, the shape of three types of gradients coils. The z-direction gradient coils are known as Maxwell pair configurations, while the x and y gradients are saddle coil configurations.

Language and Scaffolding Bubble
GRADIENT [9]:
A: an inclined part of a road or railway.
B: rate of change with respect to distance of a quantity (i.e. slope).
In MRI, the word gradient means the change of the main magnetic field (B_0) over a distance, i.e., $\Delta B/\Delta$distance that is produced by scanner hardware [10] as seen in figure 2.18.

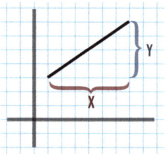

Gradient Definition Role of 'Gradients' in MRI

Figure 2.18. (Top) a gradient (incline/slope) of road as per the definition, copyright Albert Bridge, CC BY-SA 2.0. (Bottom) the gradient in MRI is the change in magnetic field divided by the change in space [8].

Recall the Larmor Frequency equation from the previous section:

$$\omega = \gamma B \qquad (2.2)$$

If you were to plot the values we just mentioned into a graph, you would see figure 2.19. The key concept here is that with a spatial gradient, a frequency term arises from the multiplication of the gradient times the location at space. For those who are ambitious, please try the thought experiments and mathematics in the appendix 2.A.1 and 2.A.2 at the end of this chapter.

But, for those who just want to get the concept and 'proceed to the conceptual intuition, below is a recipe:

1. Start with $\omega = \gamma B$, which is a constant, and B is the field the spin is exposed to.

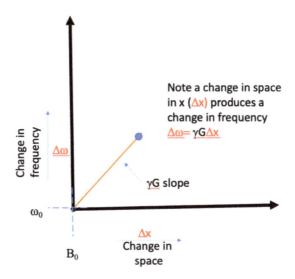

Figure 2.19. Please note the linear relationship between changes in Larmor frequency with respect to magnetic field. You might notice the following insights: (1) a change in field will result in a change in frequency, i.e., ΔB that results in a linear change in $\Delta \omega$; and (2) a change in frequency (ω) would cause a linear change in field B (related to $1/\gamma$) and the gradient. The graph is related to the calculations, as shown in the work-through example with numbers in appendix 2.A.

2. Now include a spatial gradient **G**. The gradient is a slope, i.e., has units of B field/distance.
3. Notice how we will have a different frequency at each spatial location. Put the concepts of 1 and 2 together, which gives the following equation: $\omega = \gamma(\boldsymbol{B_0} + \boldsymbol{Gx})$[15].
4. Now, now we can see that if we write $\omega_0 = \gamma \boldsymbol{B_0}$, which is a constant.
5. Then, $\omega = \gamma(B_0 + Gx)$ becomes $\omega = \omega_0 + \gamma \boldsymbol{Gx}$. By substituting $\omega_0 = \gamma \boldsymbol{B_0}$ into the equation.
6. Finally, we can see that a frequency offset term $\Delta \omega$ is proportional to γGx, i.e., $\Delta \omega = \omega - \omega_0$ and thus $\Delta \omega = \gamma Gx$.
7. $\Delta \omega = \gamma \boldsymbol{Gx}$ is plotted in figure 2.19. Imagine we are focused on this frequency offset. The frequencies range that is around the Larmor frequency is the bandwidth of frequencies we are focusing on. Note we subtracted ω_0 off the equation in 6, i.e., $\Delta \omega$ is a frequency offset that is scaled by the gradient G times the gyromagnetic ration(γ) times the distance that you would be interested in measuring). Thus, the spins at that distance 'x' have their own frequency channel! If you want to work this out, I encourage you to go through the math challenge in the appendix of the chapter[16].
8. Regardless of whether or not you are not math-oriented, perhaps this list of concepts (1-7) may be enough for you to start understanding how frequency

---

[15] A spin at position x, will see a field B that is the sum of the frequency of the main magnet (i.e., B_0) and the field that has been added by the slope of the field over a distance (a gradient). The gradient will be multiplied by the spatial position to return a factor that is a frequency dependent on the gyromagnetic ration.

[16] A concrete example of numbers of a range is 120 kHz. In MRI, an area of interest that could be used that bandwidth can be worked out to be around the operating frequency of a 1.5 T machine and by doing the calculation [63.87 MHz − 60kHz, 63.87 MHz + 60 kHz]. This would simplify to a range of [63.81 MHz, 63.93 MHz].

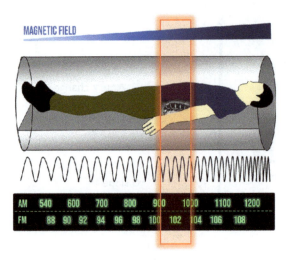

Figure 2.20. Illustrates selecting a location in the body using a gradient. Note that in this picture we are showing the bandwidth of frequencies that are mapped to a field of view or length of the region scanned.

encoding works, and you can then appreciate how a spatial position can be mapped into a frequency (offset) through the bolded equation in number 7 above. A graphical rendering that some readers may find useful is shown in figure 2.19[17].

At this point, you might recognize that there is a linear relationship of frequency with that of space (proceed to look at figure 2.20 below for more visualization of this topic). If you have made this connection, now realize that we've mapped frequencies with space, which is a powerful tool for MRI!

Gail H. (neuroscience postdoctoral researcher) reflects on how gradients are used to encode at each point in space:
"Admittedly, I was starting to get a bit uncertain about the math part of this section at first—it's been a while since I had to use some of these concepts! Then, when I considered it some more, I realized that the most important part is what the mathematical relationship means: the gradient determines the signal frequency in a distance-dependent way. So, you can take the signal you get and decode where it came from based its frequency. It's like having each location having a tag (the frequency). The 'tagged' signal is also able to carry other information, like how at a given location the number of spinning particles changes the strength of the signal (the amplitude). This is critical to make an image!"

Note that you can select a bandwidth of frequencies (over the volume) to gather in your acquisition a spatial region in the body (black lines labelled lower to upper range of where you are scanning). You can then select a single 'pixel' of space in which you want to isolate the amount of MRI signal (shown in red). Note that the gradient

---

[17] For the inquisitive reader, in appendix 2.A.1 we provide a visual derivation through graphs to illustrate the relationship between frequency offset and spatial position leading to the graph shown in figure 2.19.

'maps' the spatial value to a frequency by the collection of frequencies that are 'synthesized' as a signal. We then use a mathematical method to extract all the signals. If we know the gradient, we can map frequency terms (their amplitudes) back to the original spatial location as portrayed in figure 2.20 (which would generate a 1D image in this case). Please note that gradients can be played out in multiple directions (x, y, z) for both gradient and hardware (saddle coils in x and saddle coils in y, Maxwell pairs for z), which permits slices to be acquired in any orientation as seen in figure 2.21.

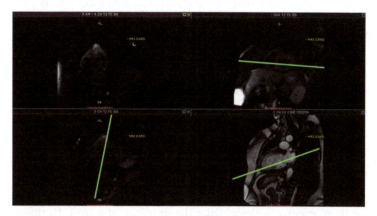

Figure 2.21. Note that the acquisition plane can be oriented in any direction (even obliquely). This is because we have the ability to acquire images in any plane. Those shown above are from a cardiac MRI exam.

History bubble

Paul Christian Lauterbur (1929–2007)[18] was a chemist who in the 1970s came up with the techniques of applying the gradient to generate an image. Prior to that, scientists only had the use of NMR spectroscopy for interrogating living matter with

---

[18] This 'File:Paul Lauterbur 2003 cropped.jpg' image has been obtained by the author from the Wikimedia website, where it is stated to have been released into the public domain. It is included within this article on that basis.

magnetic waves. The method was further improved by Peter Mansfield. In 2003, Lauterbur and Mansfield shared the Nobel Prize in Physiology and Medicine for their seminal work in MRI. His signature is shown below[19].

It is useful to see a little bit of history[20].

Lauterbur's paper was initially rejected by the journal editors of Nature. The Nature editors pointed out that the pictures accompanying the paper were too fuzzy [11]. Lauterbur, however, was able to get a rereview and it was eventually published in his seminal publication in 1973 [12].

Later, Peter Mansfield of the University of Nottingham took Lauterbur's initial work another step further by adding used frequency and phase encoding by spatial gradients of magnetic field. Mansfield developed the Fourier transformation technique, which used to recover the desired image that sped up the imaging process.

2.4.1 Summary

The key point is a given relationship between spatial position and frequency. (A linear relationship would be convenient.) Then, if you can extract the amplitudes of the frequencies, you will get the amplitude of signal at each spatial location. The word gradient means 'a slope' (like the gradient of a hill). In this case, it is the slope of B field, i.e., $\Delta B/\Delta$(distance) that is produced by hardware.

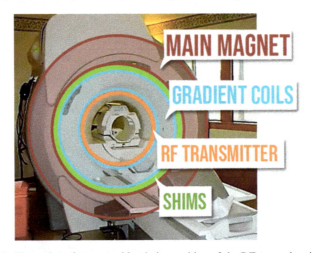

Figure 2.22. Illustration of magnet with relative position of the RF transmit coil (orange).

---

[19] This File:PaulLauterburSignature.png image has been obtained by the author from the Wikimedia website where it was made available by User:Jxramos under a CC BY-SA 4.0 licence. It is included within this article on that basis. It is attributed to Paul Lauterbur.

[20] The author was able to attend a small lecture in late 1990s when he visited with Dr Paul Lauterbur hosted at their shared alma mater (Case Western Reserve University). Dr Lauterbur relayed an entertaining story about the day he made the seminal insights into spatial encoding and MRI. Dr Lauterbur, along with Peter Mansfield, was awarded the Nobel Prize in the fall of 2003.

An additional concept is that there are coils that can drive a gradient in any direction x, y, z or any direction in space. This permits MRI 'planes/slices' to be acquired in any given direction. This coincides with the main understanding to be derived from this section: that the role of **'gradients' is to provide spatial localization in MRI**.

2.5 Radiofrequency (RF) transmitter

2.5.1 Roles of the RF transmitter
The RF transmitter provides the tipping energy needed to initiate the precession in coordination with the main magnetic field.

1) **RF pulse tips the spins**. To initiate precession, the spinning protons must be tipped, as illustrated in figure 2.24a.
2) RF transmitter utilizes $\Delta E \rightarrow M$. A changing electric current induces a temporary magnetic pulse that will tip the spins.
3) The angle to which the spins are 'tipped' is called the flip angle, as shown in figure 2.24B.

You already have two environmental factors to induce precession—the main magnetic field and the intrinsic spin of the proton (figure 2.22). The RF transmitter adds the final piece by creating a magnetic pulse to tip the protons. The magnetic field from the RF transmitter is called B1. A table from chapter 1 comparing the precession of a gyroscope to the precession of spinning protons is repeated here for convenience (table 2.2).

Table 2.2. Perturb the system and initiate the precession in conjunction with the environment set up by the main field. Recall that it creates the alignment and provides the uniform 'gravity' term.

Mechanical model of precession	MRI/NMR proton in precession
ex. Spinning top	ex. Intrinsic spin of the 'proton'
Angular momentum from mass rotating	Angular momentum intrinsic to atomic particles (the proton), intrinsic magnetic moment
Gravitation field	Magnetic field (main field) has to be very strong to get number of spins to be converted to signal
Tilting—gyroscopes are manually tipped. Perturbing the orientation of the 'top' is important to the process of precession.	*Tilting magnetic moment. But what is going to be able to tip it?* ←**NOW you need the item that is described in this section (RF transmitter)**

Carlie (chemistry/psychology/premed and dancer) reflects about transmission:

"In order to transmit efficiently in MRI, we have to operate at the Larmor frequency. It's hard to imagine how the hardware is able to do that, but I think this is the primary job of the RF transmitter. From what I understand, the RF field is transmitted at the Larmor frequency and tips the spins of the protons perpendicularly to the main field. I'll have to read on to learn more about RF transmission, but I think of the Larmor frequency as the optimal frequency at which the MRI 'communicates'."

Dr. Wu added:

"Carlie, I think you are on the right track. Understanding why it's important to center the RF transmission to be effective at specific frequency may at first feel like an abstract concept. However, to make an efficient transfer of energy, you need to be rotating that transmission at the Larmor frequency so that you are transmitting near or 'on resonance.' Basically, since the protons are precessing, if you were able to be in the same frame of reference (i.e., rotating at the same speed) you can transfer the most energy directly."

Summary
- B1 field is like an instantaneous (brief in time) 'intense' magnetic field. Note that it is perpendicular to the Z-direction (B_0 field), but rotating at the same rate as the processing dipole.

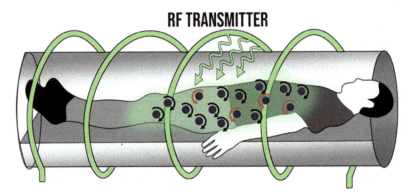

Figure 2.23. Illustration of the tipping caused by the RF transmitter. It works by changing an electrical field and causing a magnetic pulse; the magnetic pulse tips the spins to an angle (flip angle). Note that the 2D sketch illustrates the RF transmission that is typically time to occur at rates relative to the Larmor frequency (i.e., represented by the corkscrew). The transmission sends the magnetic waves that are optimized to tip the spins.

- This would cause a rotational action (RF transmitter's job is to tilt for precession) [13] in the presence of a magnetic field

If we remove the labeling, can you identify the different coils that would be in the system? Can you imagine which are outside and inside of the system? There are many more loops for the main field (bulk of the size of the MRI system) than there are for the other coils (gradient and RF transmitter). The scheme for labeling scanner components is shown in the DLS in figure 2.22.

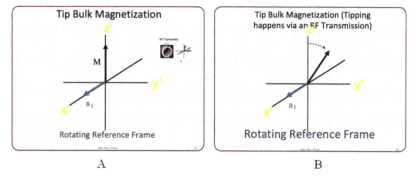

Figure 2.24. This is an illustration (you are applying a rotating field with precession frequency). Note that the amount of tipping is called the flip angle in MRI. Transmission is applied around the precessional frequency (rotating) and a magnetic pulse induces the spins to tip. B1 is the notation that describes the RF Transmitter field. B1 operates orthogonally to B_0 direction, but electrically rotates at the Larmor frequency.

Figure 2.25. Figure as a Deductive Learning Sketch for identifying parts of a magnet. Can you identify the locations as shown by the colors and which hardware is typically at these levels in the magnet?

2.6 RF receiver

Some things to observe:

1. Note that for the receiver, magnetic field lines go through a coil that induces a current $\Delta M \rightarrow E$ coming from the body (precessing spins) that goes through the receiver coil loops (figure 2.26). This is the signal that you will use to create the image.

2. Because the protons are spinning, or oscillating, the signal is like a sinusoidal wave. We only showed one frequency for simplicity. However, in reality there will be multiple frequencies over a bandwidth. This receives bandwidth around the Larmor frequency, which you must demodulate, i.e., subtract the $\gamma * B_0$.

3. You are actually 'listening'[21] to the total resultant sum of signals at a time (remember synthesis of waves). The waves of different frequencies are mixed/added together, so you need a tool to extract spatial points (see section 1.2.1.0 Fourier transform of chapter 1 for additional ideas). We will also discuss this in the last subsection of this chapter.

4. For RF receivers, the proximity to the subject to be imaged matters. The main concept is that you are letting in less noise from regions outside of your interest as it focuses on the signal.

5. Coils are often designed with a specific body part in mind so that they are designed to improve the signal from the protons to noise in their response, as described more in chapter 6 (figures 2.27 and 2.28).

Note, we've also enjoyed the diagrams of MRI hardware and descriptions that has been rendered in MRI from Picture to Proton by McRobbie, which is helpful and well done [14].

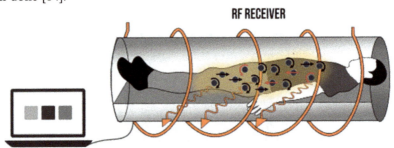

Figure 2.26. Protons will return to the main magnetic field after being tipped by the RF transmitter, releasing the energy originally absorbed. Each voxel will have a unique frequency of energy released. This signal is picked up by the receiver coil and demodulated by a computer to form an image. While this is quite difficult to render in a 2D sketch, we are illustrating the listening by the receiver to occur relative to the Larmor frequency (i.e., represented by the corkscrew). The receiver processed the magnetic waves and then sends this signal for processing.

---

[21] 'Listening' is when the receiver is turned on so signal can be received from the sample.

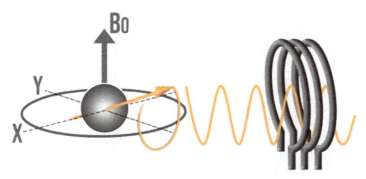

Figure 2.27. Representation of the RF receiver in the chain of signal acquisition. In this figure note the oscillatory behavior of the received wave of the signal to the RF receive coil.

RF receive coils are typically optimized in design so that these coils have the ability for fitting to patient habitus, such as size, and shape that is directed to optimizing the signal for each part of the body. This is why there are many sizes of receiver coils, as shown in figure 2.28. While these topics can require detailed discussion, we will focus on a few basic underpinnings of RF receiver concepts. For greater detail see [14, 20] and/or [21]. Additionally, radiofrequency wave representation has been discussed by Hoult [22] and others with greater attention to modeling the correct physics but for general practitioners this may be at first hard to gain fluency. For greater intuition concerning the quantum description of NMR signal and RF frequency please consider looking at Dr Hoult's work [23, 24].

For a basic and fundamental first approach in receiver coils there are two groupings of coils: volume coils and surface coils (which include phased array coils). An example of volume coil is the quadrature knee and an older style, quadrature head coils, that have a birdcage design [21]. The other main class of RF receiver coils include surface and phased array coils. The phased array coils use multiple surface coils to achieve high SNR and sensitivity to imaging an area of interest by combining signal. Because they are shaped to the portion of the body of interest these coils improve SNR by reducing added noise from areas outside that region [20, 25][22].

---

[22] Another aspect is that the multiplicity of different coils can also locally improve the SNR. Technologists must become well acquainted with working with these coils so they achieve high image quality. Physicians may also be interested in understanding which coils appear to provide them with the optimal local and feel they need for interpreting images (in terms of SNR and homogeneity for example).

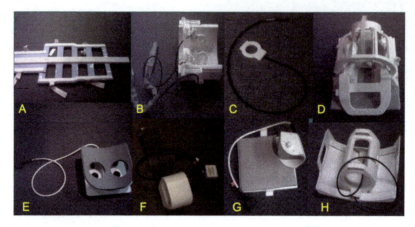

Figure 2.28. A receiver coil has best benefits when it is fitted around the area of interest. The idea is that this focuses the signal measurement to the body area, while minimizing noise coming from areas outside of the area of interest. There are head and neck coils, shoulder coils, and knee coils, as shown in this picture.

2.7 Scaffolding bubble

Interlude: behind the scenes: computers and mathematics

In this section, we discuss a few more concepts: (1) demodulation; (2) the role of the Fourier transform (the reader saw this first in chapter 1, but it is repeated here as you now can imagine the aggregate frequencies that come from the RF receiver needed for extraction into spatial components); and (3) the idea of traversing through k-space with gradients. To keep the concepts simple, we have intentionally tried to keep the concept of gradients to be simply associated with spatial localization as a point of memory[23].

1) Demodulation—In MRI, the signal is received at a bandwidth around the resonance frequency. Recall $\omega = \gamma (B_0 + Gx) = \gamma B_0 + \gamma Gx$, where γ is the gyromagnetic ratio. When receiving, we are interested in the $\Delta\omega$ of bandwidth surrounding the resonance frequency. This is called demodulation. Sorting out this huge array of overlapping and interfering signals might seem impossible, but there are signal processing ways to achieve the desired result (see mriquestions [15]). Remember that we subtracted the γB_0 term from the problem with the gradient table, as it was a constant. It turns out that we do not have to sample way up at high frequency. We take the signal from around that frequency and bring the signal down to the kilohertz range for convenience.

2) Fourier transform makes the frequency spectrum (acquired as frequency)→ spatial (as we described earlier in chapter 1).

If you understand that we are acquiring frequency terms, the multidimensional frequency space we acquire is known as k-space, and we have the paths

---

[23] At this point, simply associating the gradient with spatial encoding is sufficient. In section 2.8, we will also consider the role of gradients in terms of traversal through k-space, which we also describe in chapter five section 5.2, where we add more details on k-space after the reader has completed learning about pulse sequence diagrams.

where the spatial gradients are played out—so we can fill in the values. We can then take all the values and put them through the Fourier Transform engine that we talked about in chapter 1, and reconstruct a signal. Perhaps the figure that we presented at that point combined with your knowledge of the gradients encoding may help you now understand how the spatial encoding and reconstructing this signal works through a Fourier transform method as illustrated in figure 2.29.

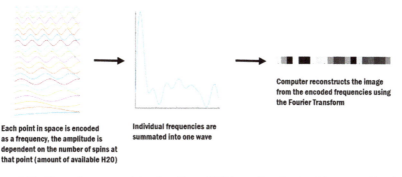

Figure 2.29. Illustration summarizing the effect of MRI encoding the signal into a combination of waves with different amplitudes. A Fourier transform (inverse) can be used to extract the original amplitudes with location. Readers with a computer science or technical background may like the reference on the FFT, which provides more details on the Fourier transform [14].

Nick P. (third-year medical student) reflected on spatial encoding with gradient:
"This was a complex concept to wrap my head around. It seems like the machine uses a constant change in magnetic field over a distance (which is the gradient) and it uses that constant gradient along with the Larmor equation as a base to map the change in frequencies from the Larmor frequency with respect to the point in space. All of these waves at their respective frequencies are in the k-space as a synthesized wave, which then is mapped as an amplitude by the inverse Fourier transform equation. These various amplitudes at their specific locations are what correspond to the changing intensities that are shown on the MRI image."

Dr. Wu said
"I like what you said, and here's an abbreviated way to think about it, if that helps:

1) Each point gets encoded as a different frequency with the amplitude being the amount of spins (amount of H_2O).
2) Listening to this signal, the images are added/summed together.
3) By the Fourier transform, we are able to unscramble those signals and retrieve the values."

2.8 The briefest of introductions to *k*-space (traversing frequency space)

The goal of pulse sequence is to **fill up *k*-space** (further discussed in chapter 5). *K*-space is frequency space. Recall chapter 1, section 1.2.1.0 (Fourier transformation). First, we have to learn what a pulse sequence is (chapter 4; figure 2.26)[24].

 These gradients are what is 'played out,'[25] along with the RF pulses by the pulse sequences, to create the images in MRI.

 Dr. Justin N. (Radiology attending) reflects on why *k*-space is important:
"I will be honest, I am no expert on *k*-space, but understanding that it exists as a repository for raw data to be converted into an image is important. Even though all other parameters may be the same for a sequence, altering how *k*-space data is filled can alter the final resulting image. These tweaks can be very helpful in certain clinical settings or when dealing with artifacts."

Applied bubble: MRI QA and medical physics

Role of a medical physicist
The role of the medical physicist in regards to MRI can be broken down into multiple responsibilities that include periodic work such as QA testing, assisting as needed, and continuous support. Collaboration and communication are vital between physicians, technologists, medical physicists, and researchers, and can benefit the organization. A key goal of the clinic is to develop strategies to obtain better image consistency and efficiency and to develop plans for maintaining good practices, including MRI safety.
  Quality assurance includes, in response to needs of the clinic:
  A. General:
    1) Gathering data;
    2) Analyzing cause of problems;
    3) Proposing solutions;
    4) Testing on a small scale;
    5) Evaluating solution;
    6) Working with vendors;
    7) Education;
    8) Writing reports to inform administration, technologists, and physicians.

---

[24] Note that the relationship between pulse sequence diagram and *k*-space will also be further discussed in chapter 5 and so the contents of chapter 4 play a central role from generating the image contrast, creating tradeoffs in SNR and time as well as influencing the impact of artifacts, which are the subject of chapters 3, 5, 6 and 7 of this book.

[25] Played out' is parlance used by MRI physicists/scientists to describe the timing and amplitudes of how the multiple gradients are turned on and off in a pulse sequence in the frequency, phase, slice directions (preview for 'directions' terms, which are covered in greater detail in chapter 3 when pulse sequences are introduced in more detail).

B. For MRI QA, specifically, the coverage of tests includes:
  1. Geometric accuracy;
  2. High-contrast spatial resolution;
  3. Slice thickness accuracy;
  4. Slice position accuracy;
  5. Image intensity uniformity;
  6. Percent-signal ghosting;
  7. Low-contrast object detectability;
  8. Artifact evaluation.
C. Periodic or intermittent for the first four items, and continuous for those after:
  1. Annual testing;
  2. Acceptance testing;
  3. Check Q/A logs;
  4. Evaluate vendor proposals for purchase;
  5. Assist with the development of safety Culture;
  6. Provide education on technology and protocols;
  7. Develop technical strategies, along with team;
  8. Assist with identifying technical issues as early as possible.

Establishing quality assurance is as much about conducting testing as it is about identifying and supporting communication and improvements to efficient solutions and ensuring safe practices.

Table 2.3. A review and expansion of the formula used for spatial encoding by a gradient.

Understanding how the gradient works through a thought exercise	A simplified derivation:
The particular 'gradient' we are thinking of in MRI is the change in magnetic field divided by the change in space. The direction of gradient can occur in any direction (i.e., x, y, z) or any combination. For a given constant gradient, consider the change in frequency as per below. Please examine the following calculations in the inset. This results in the following relationship that you will need for the next DLS. Realize that if you have a gradient in the field, you get an additive term (slope * distance) $\omega = \gamma B$ $\omega = \gamma(B_0 + \Delta B)$ $\omega = \gamma(B_0 + (\Delta B/\Delta x) * \Delta x)$ $\omega = \gamma(B_0 + G_* \Delta x)$	1. Larmor equation 2. There will be some offset to main field due to presence of other terms (i.e., gradient terms) 3. Note position offset, and then use the slope formula to add the change in B field due to the offset (i.e., $\Delta B = (\Delta B/\Delta x)_* \Delta x$) 4. Substitute $G = (\Delta B/\Delta x)$

(*Continued*)

Table 2.3. (*Continued*)

Understanding how the gradient works through a thought exercise	A simplified derivation:
	that is added to the frequency $\omega = \gamma(B_0 + (G_* \Delta x))$ Taking this further as an exercise with the following directions: 1. Plug in the value for Δx in column two to find an expression for ω 2. In column 3, calculate the offset frequency $\Delta \omega$ (relative to Larmor) frequency by subtracting $\gamma * B_0$ 3. For this step, assume a fixed gradient ($G = 10$ Gauss cm^{-1}) and plug that value as well as the value for Δx for column 4 to continue to calculate $\Delta \omega$ 4. With the given information you have already $G = 10$ Gauss cm^{-1}, Δx, insert the value for $\gamma = 42.57$ MHz T^{-1}. Note that MHz = 1e6 Hz, kHz = 1000 Hz, 1 meter = 100 cm, and that 1 Gauss = 10^{-4} Tesla.

We now hope that you **make the connection** that each spatial position is encoding a frequency (as in column 4) **and that you see there is a linear relationship between changes in frequency ($\Delta \omega$) to Δx**. If you are having trouble with this concept or relationship, I think you can safely assume that the machine is encoding the spins in location through the gradient and realize that it could aid your conceptualization of MRI, but you can also move forward to later concepts (table 2.4).

Gail H. (neuroscience postdoctoral researcher) reflects on chapter two:

"To me, the overall takeaway is that the MRI scanner is comprised of layers nested inside of each other, and each layer is important for you to get an image. First, you need a huge magnetic field to get the protons aligned and 'ready' in the person's body. The shims help to make the main magnetic field more uniform over the entire body, so the image won't be messed up. The gradient coils provide a way to make the magnetic field vary in precise ways. The gradients are important for spatial localization. You need them for spatial information to be eventually decoded when you do a scan. You also need gradients to help select what slice of tissue to look at (again, having to do with something spatial). Next, you need an RF transmitter to tip the protons. As the protons go back to their starting position, they release energy. You need radiofrequency receiver coils to record that energy being released. The data can then be decoded to get an image. Now that I understand how the MRI scanner layers contribute to making an image, the process makes a lot more sense!"

Table 2.4. Please fill in the question marks in this exercise. This table illustrates how spatial position can be $\Delta\omega$ (frequency as expressed with difference to the resonance frequency). Note that the signal processing to observe a bandwidth that surrounds the resonance frequency is called demodulation (the demodulation concept is further described in section E). Note that the goal is to understand that it is possible to have points in space (Δx) to be encoded as a different frequency ($\Delta\omega$). *Extra Credit* — if you have a 300-cm object, how much frequency would you need to cover that object?

Spatial position	ω	$\Delta\omega$	$\Delta\omega$ for the given values below	$\Delta\omega$
Fill in numbers for spatial position in cm.	Find ω $\omega = \gamma B$ at Δx	Find $\Delta\omega = \omega_0 - \gamma^* B_0$ ($\Delta\omega$ with subtraction of the Larmour Frequency)	Given $G = 10$ Gauss cm^{-1} Δx is known from column 1	Given $G = 10$ Gauss cm^{-1} Δx is known from column 1 Gyromagnetic ratio is given
Δx (generalized)	$\omega = \gamma^*(B_0 + G\Delta x)$ Substitute for Δx	$\Delta\omega = \gamma\, G\, \Delta x$ Δx is known Subtract $\gamma^* B_0$ From ω	$\Delta\omega = \gamma\, G\, \Delta x$ (G and Δx are known)	$\Delta\omega =$ (with G, Δx, γ known)
1 cm	$\gamma^*(B_0 + G)$	γG	?	4.26 kHz
2 cm	?	?	?	8.52 kHz
3 cm	?	?	?	?
4 cm	?	?	$40\,\gamma$	
...			...	...
Ex. $G = 10$ Gauss cm^{-1} 2 cm G (2 cm) = 20 Gauss*	Note $B = B_0 + Gx$ $B =$ 42.57*(1.5 T + Gx) 42.57 MHz T^{-1} (1.5 T +20 Gauss*) *Using 2 cm point to the left	Note that the terms Increase linearly With a slope of $10\,\gamma$ γ is a constant $\omega = \gamma B$		Note use $\gamma = 42.57$ MHz T^{-1} 1 G cm^{-1} = 1e-4 T cm^{-1} And MHz = 1e6 Hz kHz = 1000 Hz m = 100 cm
	Find ω $\omega = \gamma B$ at Δx	Find $\Delta\omega = \omega_0 - \gamma^* B_0$	Given $G = 10$ Gauss cm^{-1}	

(*Continued*)

Table 2.4. (Continued)

Spatial position	ω	$\Delta\omega$	$\Delta\omega$ for the given values below	$\Delta\omega$
Fill in numbers for spatial position in cm.		($\Delta\omega$ with subtraction off the Larmor frequency)	Δx is known from column 1	Given $G = 10$ Gauss cm^{-1} Δx is known from column 1 Gyromagnetic ratio given
Δx (generalized)	$\omega = \gamma^*(B_0 + G\Delta x)$ Substitute for Δx	$\Delta\omega = \gamma\, G\, \Delta x$ Δx is known Subtract $\gamma^* B_0$ From ω	$\Delta\omega = \gamma\, G\, \Delta x$ (G and Δx are known)	$\Delta\omega =$ (with G, Δx, γ known)
1 cm	$\gamma^*(B_0 + G)$	$\gamma\, G$ (cm)	10 γ (Gauss)	4.26 kHz
2 cm	$\gamma^*(B_0 + 2G)$	$2\gamma G$ (cm)	20 γ (Gauss)	8.52 kHz
3 cm	$\gamma^*(B_0 + 3G)$	$3\gamma G$ (cm)	30 γ (Gauss)	12.78 kHz
4 cm	$\gamma^*(B_0 + 4G)$	$4\gamma G$ (cm)	40 γ (Gauss)	17.04 kHz
...			...	
Example $G = 10$ Gauss cm^{-1} 2 cm $G(2\text{ cm}) = 20$ Gauss*	Note $B = B_0 + Gx$ $B = 42.57*(1.5\text{ T} + Gx)$ 42.57 MHz T^{-1} (1.5 T + 20 Gauss*)*Using 2 cm point to the left		Note that the terms increase linearly with a slope of 10 γ γ is a constant	Note $\gamma = 42.57$ MHz T^{-1} MHz = 1e6 Hz kHz = 1000 Hz

2.9 Summary and looking ahead

We went over several concepts in this chapter. Hardware can set the stage to reflect on the inner workings of the scanner to generate a deeper understanding of MRI. Here is a checklist summary of the main ideas.

   I. Main field
      1. Aligns spins.
      2. Permits precession.
      3. Sets the resonance frequency.
   II. Shim
      1. Achieves flat magnetic field (remove unintended magnetic gradient fields).
      2. Understand the role of the solenoid in the main field.
   III. Gradient
      1. Provides spatial encoding.
      2. Specialized coils can be applied in multiple directions.
   IV. Radiofrequency transmitter
      1. Tips 'spins', which are needed for precession. (Note that you also need the main field to permit precession, i.e., you need both the main field and the RF transmitter.)
      2. RF here is generated by a quick burst of magnetic energy at resonance frequency.
   V. Radiofrequency receiver coils
      1. Listen to signal an RF coil.
      2. Select a bandwidth to listen.
   VI. Behind the Scenes
      1. Demodulation makes receiver processing easier, as it removes the offset constant Larmor frequency term ($\gamma^* B_0$) to just yield the offset, which is used for final spatial localization.
      2. Fourier transform is the technique that converts frequency encoded images and turns them into their spatially reconstructed images.
      3. Traversal through k-space (acquired frequency space) is related to the pattern of the gradients on the pulse sequence diagram (discussed more in chapter 5).

Appendix 2.A

For many learners, it is helpful to understand how encoding of signal at locations is encoded into waves. This uses the **Larmor** and **gradient** relationships we previously discussed. We pictorially explain the relationship below, which may be useful for technologists and medical professionals. Additional information is also contained in this appendix (of chapter 2) for readers who are even more inclined to learn the concept by looking at formulas as an alternative way to learn the concepts.

We acknowledge that for some readers that there is an 'abstract' connection between frequency encoding and space. Initially, this may feel a little unapproachable, and is quite normal, as it was only in the year 1971 that Paul Lauterbur jotted notes on this relationship that permitted spatial encoding of MRI images despite that nuclear magnetic resonance (NMR) discovery was nearly 25 years earlier [17]. His original notebook is shown in this image [18]. Later, Peter Mansfield introduced the Fourier transform technique that made MRI begin to be realizable for hospitals and clinics around the world [19].

However, we have found that for some people, the materials of these additional sections 2.A.1 and 2.A.2 ended up helping them solidify their conceptual understanding.

The key point is to see that starting out with two relations that:

$$\text{a) Larmor relationship: } \omega = \gamma B \qquad (2.3)$$

$$\text{b) Spatial gradient relationship: } G = (\Delta B / \Delta x) \qquad (2.4)$$

When (3) & (4) are combined and with a little algebra of multiplication, slope manipulation, and subtraction that you get a final relationship

$$\Delta \omega = \text{constant} * \Delta x \qquad (2.5)$$

Note that the constant is equal to γG *because it is constructed from the multiplication of two constants (γ is a constant for water and that G is a constant[26])*

Here's a recipe:
1. *Then the reader should imagine that at each space there is signal content (for example, number of hydrogen spins available for that water signal is the signal content).*
2. *Next, imagine that at each point you've mapped at space to a frequency. That frequency is carrying that 'amplitude' of signal by a wave.*
3. *Because these RF signals (the frequency carrying amplitudes) are simultaneously acquired across a bandwidth of frequencies. This combined signal is a combined sum of mixed signals.*
4. *The mixed RF signals[27] are decoded by the 'Inverse Fourier Transform', which is a mathematical tool.*

---

[26] In reality, we can attempt to keep this gradient G constant as we 'listen' to the acquired signal. In fact, even if the gradient is nonlinear, there are ways to compensation for that nonlinearity and still obtain usable signal. One other note, if you understand the relationship between gradient, space and frequency, then understanding frequency encoding in section 4.2 will appear more natural and your concepts will be able to bring you to the next levels of conceptualization in MRI.

[27] It may be interesting to think of the MRI machine as itself performing a 'Forward Fourier Transform' of these signals. When the signals reach the RF receivers, it then must be decoded by the 'Inverse Fourier Transform' to return the signal to produce the images that we see in MRI.

2.A.1 A visual approach to understanding why $\Delta\omega = \text{constant} * \Delta x$.

Below is a visual way to think about the recipe for between encoding in space with frequency (figure 2.A.1a–2.A.1d).

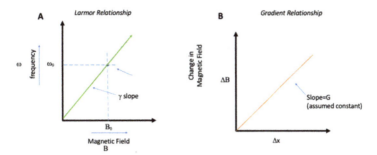

Figure 2.A.1.a. Start with the two things we already know. A) Recall Larmor's equation $\omega = \gamma B$, but look around B_0, (i.e. $\omega_0 = \gamma\, B_0$), γ is the gyromagnetic ratio. Note that ω_0 and B_0 are both constants for a known field and that we are evaluating them to be for water. B) However, if we have a way to change B using a gradient, then we can have the gradient relationships. This means that we can have a change in magnetic field B that corresponds to a change in space, x. Assume that the gradient has a constant slope (we will discuss this more in chapter 4 when we describe pulse sequence diagrams).

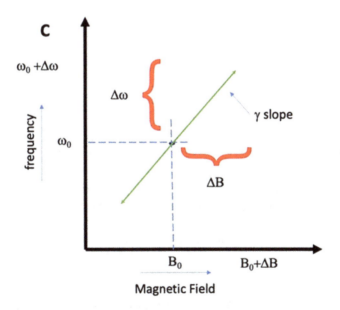

Figure 2.A.1.b. A) Note how the change in field ($\Delta B'$) creates the change in frequency ($\Delta\omega'$) will still follow Larmor's equation. Where prime is an indicates an arbitrary difference value for ΔB and $\Delta\omega$. Note that the slope is γ as it follows the Larmor equation $\omega = \gamma B$.

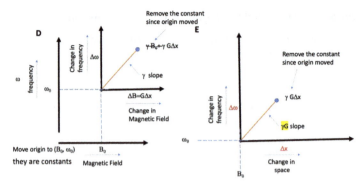

Figure 2.A.1.c. A) Here we can consider a change in magnetic field ΔB due to the gradient (we have incorporated (figure B) into our current picture. B) Move the origin from O = (0,0) to (B_0, ω_0). Instead of having the change in magnetic field be on the x-axis, let's consider the change in space and move the gradient value G to the slope of the line.

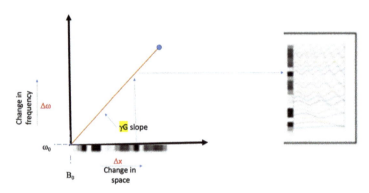

Figure 2.A.1.d. In the final figure, note how change in space is 'mapped' to a change in frequency (we used this when we talk about encoding space as a frequency and the inverse Fourier Transform).

2.A.2 An exercise to understand why $\Delta\omega$ = constant * Δx with real numbers

We provide several tables and a graph as an exercise for the reader to explore with formulas. The advantage of this section is to have concrete numbers that are in common use in scanners. *This section may be particularly useful for a few advanced residents and technologists, researchers, medical physicists, and biomedical engineers*[28].

Below is an illustration for figure 2.A.2, which illustrates numbers used in the example.

---

[28] We found that this formula was helpful for about 1/3 of our radiology residents when we discussed it in person. So, we feel that it is better as a handout and a walkthrough for those who may want to stay behind and go through this exercise. It is, however, useful for medical physicists and biomedical engineers and technologists who may want to have a larger clinical role in the education and conceptual understanding of MRI.

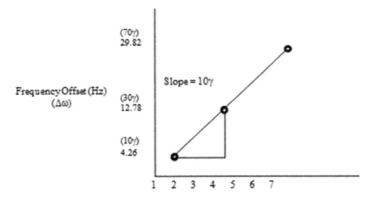

Figure 2.A.2. Please note the linear relationship between changes in Larmor frequency with respect to magnetic field. You might notice the following insights: (1) a change in field will result in a change in frequency, i.e., ΔB that results in a linear change in $\Delta \omega$; and (2) a change in frequency (ω) would cause a linear change in field B (related to $1/\gamma$) and G. Real world numbers in MRI are placed on the linear graph to shown with the numbers used in the exercise in tables 2.2 and 2.3.

References

[1] Koch K M, Rothman D L and de Graaf R A 2008 Optimization of static magnetic field homogeneity in the human and animal brain in vivo *Prog. Nucl. Magn. Reson. Spect.* **54** 69–96
Togo H, Rokicki J, Yoshinaga K, Hisatsune T, Matsuda H, Haga N and Hanakawa T 2017 Effects of field-map distortion correction on resting state functional connectivity MRI *Front. Neurosci.* **11** 656

[2] Khan Academy (nd) Magnetic fields through solenoids (video) (Retrieved April 6, 2022) https://www.khanacademy.org/science/in-in-class10th-physics/in-in-magnetic-effects-of-electric-current/magnetic-field-dueto-current-carrying-loops-and-solenoids/v/magnetic-fields-through-solenoids

[3] IMAIOS (nd) Nuclear spin and MRI (Retrieved April 6, 2022) https://www.imaios.com/en/e-Courses/e-MRI/NMR/nuclear-spin

[4] Nuclear precession (n.d.) Questions and Answers in MRI (Retrieved April 6, 2000) https://mriquestions.com/why-precession.html

[5] Broadhouse K (n.d.) *The Physics of MRI and How We Use It to Reveal the Mysteries of the Mind* (Queensland: University of the Sunshine Coast) (Retrieved February 16, 2022) https://research.usc.edu.au/esploro/outputs/journalArticle/The-Physics-of-MRI-and-How/99450906702621

[6] Precession and Larmor frequency (n.d.) *IMAIOS* Retrieved April 6, 2022 https://www.imaios.com/en/e-Courses/e-MRI/NMR/Precession-and-Larmor-frequency

[7] Definition of SHIM (n.d.) Retrieved April 15, 2022 https://www.merriam-webster.com/dictionary/shim

[8] Magnetic shimming (n.d.) Questions and Answers in MRI. Retrieved February 18, 2022 http://mriquestions.com/why-shimming
[9] Definition of gradient (n.d.) *OED Online*. Retrieved April 15, 2022 https://www.oed.com/view/Entry/80428
[10] Gradients (n.d.) Questions and Answers in MRI. Retrieved April 6, 2022 http://mriquestions.com/what-is-a-gradient.html
[11] Paul Lauterbur (n.d.) *The Enonomist* Retrieved April 6, 2022 https://www.economist.com/obituary/2007/04/04/paul-lauterbur
[12] Lauterbur P C 1973 Image formation by induced local interaction; examples employing nuclear magnetic resonance *Nature* **242** 190–1
[13] B1 effect (n.d.) Questions and Answers in MRI. Retrieved January 26, 2022 http://mriquestions.com/how-does-b1-tip-m.html
[14] McRobbie D W, Moore E A, Graves M J and Prince M R 2006 *MRI from Picture to Proton* 2nd edn (Cambridge: Cambridge University Press) https://doi.org/10.1017/CBO9780511545405
[15] NMR signal (n.d.) Questions and Answers in MRI. Retrieved March 22, 2022 http://mriquestions.com/signal-squiggles.html
[16] Maklin C (n.d.) Fast Fourier transform *Towards Data Science* Retrieved March 22, 2022 https://towardsdatascience.com/fast-fourier-transform-937926e591cb
[17] Bloch F, Hansen W W and Packard 1946 Nuclear induction *Phys. Rev.* **69** 127
[18] Purcell E M, Torrey H C and Pound R V 1946 Resonance absorption by nuclear moments in a solid *Phys. Rev.* **69** 37–8 (the original announcement by Purcell's group)
[19] Rabi I I, Zacharias J R, Millman S and Kusch P 1938 A new method of measuring nuclear magnetic moment *Phys. Rev.* **53** 318 (the first demonstration of resonance absorption in LiCl, whose graph is shown above)
[20] Bushberg J T, Seibert J A, Leidholt E M Jr and Boone J MK M 2011 The Essential Physics of Medical Imaging (3rd edn) (Philadelphia, PA: Lippincott/Williams & Wilkins)
[21] Vaughn J T and Griffiths J R (ed) 2012 RF Coils for MRI (New York: Wiley)
[22] Hoult D I 2009 The origins and present status of the radio wave controversy in NMR *Conc. Magn. Reson. A* **34A** 193–216
[23] Hoult D I 2011 The principle of reciprocity *J. Magn. Reson.* **213** 344–6
[24] Hoult D I and Richards R E 2011 The signal-to-noise ratio of the nuclear magnetic resonance experiment. 1976 *J. Magn. Reson.* **213** 329–43 (https://onlinelibrary.wiley.com/doi/10.1002/cmr.a.20142)
[25] Brown R W, Cheng Y-C N, Haacke E M, Thompson M R and Venkatesan R 2014 Magnetic Resonance Imaging: Physical Principles and Sequence Design 2nd edn (New York: Wiley)

IOP Publishing

MRI: Connecting the Dots
A start to concepts
Dee Wu

Chapter 3

Basic building blocks for MRI

(Concepts in contrast resolution, relaxation, critical thinking for weighting/values, and T2*)

3.1 Introduction

Some of the enthusiastic readers, who perhaps even have prior experience, may have already leapt forward, and arrived at this chapter![1] However, whatever level of experience you as the reader are at in your MRI journey, this chapter could be a great place for you to strengthen your knowledge. Certainly, contrast and resolution are vital to a greater grasp of the basis for relaxation and image weighting (i.e., T1, T2, T2*), which have direct applicability for clinical application of MRI. Through the concepts of relaxation, contrast, and resolution, this chapter provides a great opportunity to engage and review or begin to start on major concepts of MRI. This is a simplified comparison as a house metaphor, and I'm taking a few liberties to communicate the concept that we will be finishing the outward appearance of the house, starting with the fact that we have finished the foundation of this house, as portrayed in figure 3.1.

(a). For those of you who know a little bit about construction, understanding MRI is like building a house. There are several steps, from drafting to building the skeleton, but you have now reached the chapter in chapter 2 that begins to ''frame'

---

[1] In the preface of this book, we have provided table A1 to guide readers in deciding where to begin reading the book. No matter where the reader is coming from, the content in this book is provided to introduce the core principles in a logical fashion, but the time-pressured reader may feel they want to start here in this chapter, and look back at the first ten concepts when they are ready.

Figure 3.1(a). The analogy of developing the 'foundation' when building a house that we used to describe the content expressed in chapters 1 and 2 as a process of learning MRI in this book.

Figure 3.1(b). After you have built your foundation for learning MRI. It is now time to ascend to the 'framing' of the house being an analogy for your continued progression into understanding MRI.

the house [1]. This is a simplified comparison, but I'm taking a few liberties to communicate the concept that we will be framing a house, as portrayed in figure 3.1(b).

Up to this point, we've laid the foundation from chapter 1 (Core Concepts):

1. Waves; 2. Water; 3. Multiple looks; 4. Dipoles and Precession; 5. $\Delta E \rightarrow M$, $\Delta M \rightarrow E$.
And we have the framing (hardware) in chapter 2: 1. main field; 2. RF transmitter; 3. Gradient; 4. Shims; 5. RF receiver.

Natalie said:
"I see you have two parts we have to build on chapter 1 (the 5 core concepts), which is like the foundation of the house. And, then in chapter 2, which is about the MRI hardware, which it looks like five more concepts to learn. I'm thinking that with these ten concepts between the five core concepts, as well as an overview of hardware, we will have a good start to learning MRI."

The core concepts lay a scientific 'foundation' to understand what principles MRI is built from, while the 'framing' provided the scaffold topics relating to hardware. The next step is to 'finish' the house. The singular word 'finishing' could seem to imply to some that you are mostly complete, but we hardly mean to suggest that we understand all the underlying complexity of MRI in medicine in a few chapters! In chapter 3, we will add the more substantial details that begin to make a house functional.

The concepts in this chapter determine the 'exterior design,' or the appearance of the MRI image. Let's start looking at the bricks surrounding the frame, which include:

T1 and T2 relaxation, create the **contrast resolution** in an image[2].
Pulse sequences, create the multitude of looks that MRI can accomplish.
Spin echo, is a technique that can minimize susceptibility artifacts.
The mastery of MRI may take time to develop, but the understanding of these pulse sequence 'blocks'[3] can improve the reader's

---

[2] Proton density (PD) weighted image weighting will be described later near the end of the chapter, but was not included in this comment to focus the reader on the two main contrast weightings that people mostly apply to a majority of pulse sequences in MRI.
[3] Pulse sequences diagrams (and the blocks) are described further in chapter 4.

conceptualization that is aimed at providing future opportunities that can help them advance their clinical practice. Please bear in mind that MRI signals are highly dependent on the relaxation effect (notably T1 and T2, which we will further discuss in this chapter). The MRI signal results from the fact that the signal is based not only on the hardware, but also the subject's (the patient's) properties. For imaging on patients, the nuclei observed in clinical MRI are the hydrogen nuclei, which are especially associated with water and fat.

As we previously mentioned in chapter 1, recall that in terms of abundance we are made up of mostly water. Bringing us back to other core concepts for chapter 1, also recall that we stressed the importance of 'multiple looks,' which MRI provides, and its ability to manipulate multiple relaxation contrasts. Understanding the numerous pulse sequences and being able to produce multiple contrasts in MRI can provide unique advantages to improve diagnosis. This is one of the most challenging, yet fascinating parts of MR imaging. Contrast and resolution play a key role in the ability to locate and find disease. You will balance these perspectives throughout your hopefully long and exciting career.

3.2 Contrast and spatial resolution

Learning objectives:
1) What does the word resolution mean?
2) What is contrast?
3) What is spatial resolution?

Both **contrast** and **spatial resolution** are fundamental parameters that can be used to describe image quality and assist in achieving clinical diagnoses. These ideas are cornerstones to the acquisition of good images. The concept of contrast is a gateway to additional MRI topics, such as relaxation, which are soon to be discussed as follows.

There are many strengths in MRI, as it is constructed from water, and thereby is particularly good at representing certain disease processes. Inflammation/wound healing, angiogenic processes in cancer, and pericardial effusion are indicative of strong shifts in water distributions and thus may be good candidates for clinical evaluation by MRI.

While MRI provides superior **contrast**, CT or fluoroscopy provide superior **spatial resolution.** Which is more important is dependent on the clinical application. MRI is also particularly adept at presenting **multiple looks**. Images with varying contrasts provide different perspectives that allow you to hone in on a diagnosis.

> **Word etymology**
> The following definitions are provided by Merriam-Webster [2]. Let us see how we can apply them to a clinical backdrop.
> - **Contrast** is the degree of difference between the lightest and darkest parts of a picture.
> - **Resolution (spatial)** is the ability to *distinguish* between or make independently visible adjacent parts of an object.[4]

3.2.1 Contrast

The term '**contrast**' generally refers to the ability to distinguish between two shades of materials. Contrast creates highlights, shadows, and clarity within an image. Contrast agents are added to bring out differences between different structures. For example, gadolinium-chelated agents[5] can be used to create contrast between normal tissues and areas of concern. In the clinic, it is important to be able to distinguish between images pre- and post-contrast (contrast agent injection).

Consider the differences in the shades of grey below. Group A exhibits a higher level of contrast than does group B (see figure 3.2). To **distinguish between different tissues**, we need to obtain contrast between different tissue types as well as between normal and abnormal tissues, which is indicative of disease as depicted in figures 3.3 and 3.4).

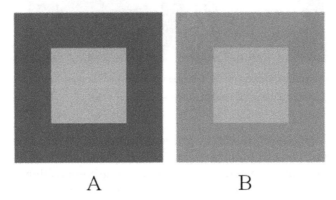

Figure 3.2. The internal squares of both group A and B are identical shades. However, because the outside shade of group A has greater contrast, the square is easier to define.

---

[4] Contrast resolution reflects the combined ability for discrimination of contrast between tissues/disease at different levels of spatial detail. The eye becomes less able to separate contrast at smaller levels of spatial dimensions. Note that the word resolution is sometimes used only to mean spatial resolution. At other times, the term can be used in the clinic/when conducting research to indicate the contrast resolution, as above.
[5] **Chelating** is used to bind metal ions, which sometimes alone are toxic to the body. By forming metal **chelates**, these ions are better tolerated and are more efficiently eliminated by the body.

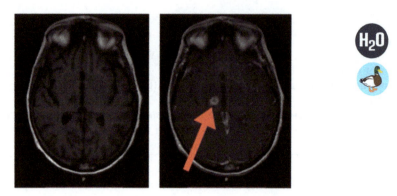

Figure 3.3. The image on the left shows low contrast of a tumor. However, after contrast agent injection, there is high contrast between the lesion and the brain in the image on the right. Note, in this case, the visualization of a tumor is often enhanced through use of the contrast agent in MRI. A note about windowing/adjusting of image: contrast is set to provide visualization of the lesion (tumor) in these images. The lesion has negative contrast (i.e., dark) in the left image (no contrast agent injection). After an MRI contrast agent has been injected, the lesion has a bright signal, which allows better visualization of the tumor and potentially the lesion's vascularity.

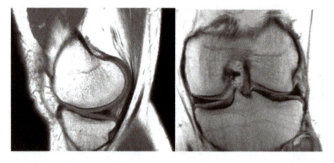

Figure 3.4. Note the contrast difference in the cartilage versus the bone in these knee images (Sagittal (left) T1 and Coronal (right) proton density [PD] weighted views)[6].

Why do we see varying shades of grayscale values in the knee on an MRI? The knee, or any joint, consists of many parts. We have bone, cartilage, muscle, tendons, arteries/veins, and more. Each tissue type has a unique composition. For example, in articular cartilage, there is water, proteoglycan, and collagen content, all of which influence the signal. Even the difference between a bound fluid and moving fluid will result in varying contrast.

The discussion on cartilage may seem niche, but provides an example to emphasize how important it is as a physician, technologist or MRI scientist to develop a strong understanding of anatomy and physiology in order to identify disease processes and progression. Greater intuition in identifying the abnormal from the normal will improve your diagnostic skills in the long run as you proceed through residency,

---

[6] Proton density (PD) weighted image. Weighting will be described at the end of the chapter.

fellowship, and beyond. In the section on relaxation, we will continue to build up steam around these issues.

Nick P. (third-year medical student) reflected on clinical importance in the previous section on contrast agents in imaging:
"Imaging with and without contrast agents is very important in finding underlying disease processes and is just another player in the 'multiple looks' that MRI is so good at supplying. Contrast agents can bring out the details of subtle lesions. It seems like what is just as important is knowing what you are trying to find in the imaging so that you can order appropriate studies to find it, since some modalities are better at picking up subtle details than others."

3.2.2 Spatial resolution

Objects are separated between different points in space, which is referred to as the resolution in MRI. **Spatial resolution** refers to the ability to differentiate two adjacent structures as being distinct from one another. Spatial resolution involves a trade-off between precision and clarity, as illustrated in figure 3.5.

Similar to the concept pictured above, MRI medical physicists evaluate resolution in an image using phantoms (figure 3.6). Phantoms are specially designed objects used in quality assurance to assess varied parameters of an MRI scanner. The phantom pictured has holes drilled at known distances. These holes can be filled with water or contrasting agents to mimic MRI conditions. For the evaluation, you must determine which value is the smallest grid that is still capable of providing clear resolution.

There are three grids of varying size, each with an upper and lower half. The upper left will be used for the horizontal resolution and the lower right for the vertical resolution. In this example, we can clearly define horizontal separation within the '1.1-mm' grid and in the '1.0-mm' grid, but not in the '0.9-mm' grid. Therefore, '1.0 mm' is our limit of resolution. Details can be found in the ACR MRI Phantom Quality Assurance Manual.

In sections 3.1.4–3.1.6, we provide several illustrative examples for the understanding of clinical objectives, including aneurysms, brain metastases, and vascular runoffs (bolus chasing), in MRI. For these examples, consider the contrast and spatial resolution required to meet the objectives of the exam.

Figure 3.5. As the lines decrease in size and become closer together, they become more difficult to distinguish from one another.

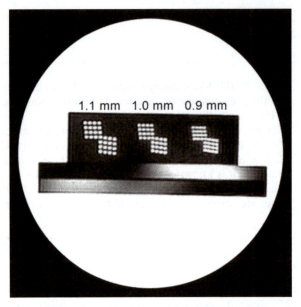

Figure 3.6. A phantom is a surrogate (i.e., plastic encasing of water) for a real-life object that helps provide quality assurance [3].

Applied bubble: clinical example

While MRI can provide great contrast, CT is often better suited for imaging blood and bone structures, particularly when more resolution is needed. In the case of a saccular aneurysm, MRI does not always provide enough resolution, as displayed in figure 3.7. A surgeon must be able to assess the width of the aneurysm neck in order to provide treatment. The aneurysm can then be closed off by filling with material (coil embolization), or clipped. Traditionally, wide necked aneurysms need to be clipped, as the coils may not stay within the aneurysm[7]. This has changed over the last several years with the advent of stent-assisted

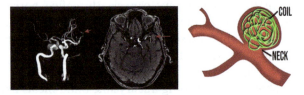

Figure 3.7. (Left) looking at a saccular brain aneurysm, the MRI does not appear to demonstrate enough structural details due to low spatial resolution. However, it has decent contrast. (Right) Saccular aneurysms are a common type of brain aneurysm with a neck that connects the aneurysm to the artery. The size of the neck influences choice of treatment.

---

[7] Many factors influence whether an aneurysm is clipped or coiled, including neck size and ease of access, or location. For example, middle cerebral arteries (MCA) bifurcation aneurysms may be easier to access for clipping. In other areas, such as the basilar tip aneurysms that are very difficult to access via a craniotomy, aneurysms are more generally coiled.

Figure 3.8. The link is to a video produced by a physician that describes some of the goals for therapeutic treatment of aneurysms and coiling and clipping of these aneurysms [3].

coiling, and the technology should be monitored to keep abreast of all changes in technology that can assist with patient care, as portrayed in the video in figure 3.8 [3].

Understanding spatial resolution can be critical in clinical care. In this application, CT and/or digital subtraction angiography have a greater ability to achieve higher spatial resolution.

Applied bubble: clinical example

MRI is preferred for brain metastases due to the need for both contrast and spatial resolution. We refer to the combination of the two as contrast resolution. The smaller the object, the greater the contrast needed to differentiate between structures. Detection of metastatic lesions, small and large, impact a patient's prognosis. The presence of asymptomatic brain metastases may alter the staging of certain malignancies, thereby changing treatment strategy. In some cases, this can change a treatment focus from being curative to one of palliative intent, which can be demonstrated in figure 3.9.

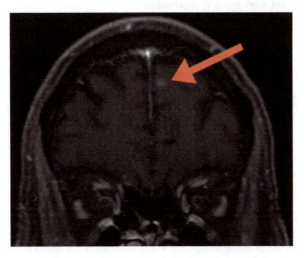

Figure 3.9. The contrast between the metastasis and the parenchyma is not as strong, which is the reason that contrast agents are used to help the visualization of these lesions.

Applied bubble: clinical example

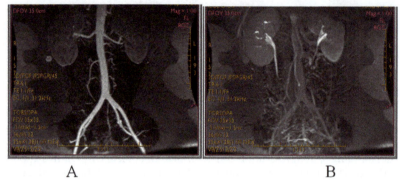

A   B

Figure 3.10. Demonstrates two phases of a 'runoff' arterial and delay phase in this image. In these images, there appears to be sufficient resolution between the blood and surrounding tissues for diagnostic purposes.

Another example when comparing MRI with CT includes angiography. MRI provides sufficient spatial resolution for diagnostic evaluation. Contrast dye can be added to further increase clarity of the blood vessels, as can be visualized in figure 3.10.

Summary on contrast and spatial resolution

Both contrast and resolution play a key role in clinical diagnoses of many images. Different tissues have different levels of contrast. One reason for different signal levels is the abundance (or lack of) protons, which would cause an alteration in contrast between the multiple tissues. The behavior of the signal generated by the magnetic dipoles would be impacted by the topic known as relaxation.

3.3 Relaxation (what, how, and when?)

Relaxation is an important concept, as it is the basis of **contrast and spatial resolution** in much of MRI. It can be challenging for beginning learners, as the ideas may feel slightly abstract, but the centrality of the understanding is key for what follows. This includes the basis of clinical interpretation of images, artifacts, and pulse sequences.

Learning objectives:
1) Beginning with a clinical point of view: what is 'weighting,' and how and which parameters (TR, TE) alter the T1 and T2 appearance.
2) Impact of regrowth and decay on appearance: what is the longitudinal direction and transverse direction; *explain T1 as regrowth, T2 as decay*, as well as the potential combination.
3) Intermediate clinical details on the appearance of weighted images: can you describe the T1 and T2 appearance of clinical images? Then, can you describe the appearance of fluid/ fat on T1 and T2 images, with and without fluid and/or fat suppression?

3.3.1 What does weighting look like?

Some people describe the general appearance of T1 and T2 on clinical images as the 'weighting' of the image. The weighting attributes to 'multiple looks.' Dr. Allen Elster (from Wake Forest University and who created www.mriquestions.com) has reasonably described the challenges of communicating T1/T2, as written below:

'…terms such as 'T1-weighted' and 'T2-weighted' are among the most overused and least clear concepts in MR imaging. In the broadest sense, these terms are used to communicate to other physicians the type of MR pulse sequence employed to generate a series of images.

…In a narrower sense, an (image weighting provides the) implication that a single intrinsic tissue parameter dominates the image contrast observed [4].'

This concept of weighting leads us back to our topic on 'multiple dimensions' from the multiple looks section in chapter 1. MRI can be an excellent problem-solving tool given the possibility of portraying multiple dimensions (dimensions are another way to describe multiple contrasts or appearances), which permits the appearances that help the physician or 'reader' to be more confident in their conclusions, as we have described before.

3.3.2 Scaffolding bubble: Synthetic imaging, creating images from weighting parameters, and synthesis

Recently, 'synthetic imaging' has been introduced to the clinical scanners [5]. The idea is that you acquire multiple weightings, and after the acquisition you have all the possibilities of different image weightings (T1, T2, FLAIR), through the rendering of the image by using the Bloch equations. The method is not without challenges, as it does not always truly create the look that physicians have come to expect (and in particular, synthetic T2-FLAIR may not always achieve desired results), so these contrast distortions can be problematic. Perhaps a future iteration of these algorithms could make a breakthrough in the ability to generate adequate and correct contrast that will become more accepted across many clinics.

3.3.3 How do TR and TE affect weighting of T1 and T2?

We are going to jump in and describe how sequence parameters, particularly repetition time (TR) and time to echo (TE), affect the 'T1-weightings' and 'T2-weightings' of an image. We will introduce the concept now, but go further in depth later when discussing pulse sequence diagrams in chapter 4.

By altering these parameters, we define the contrast 'type' of the MRI image. There are multiple types of images; T1-weighted, T2-weighted, and proton density (PD) are the three main sequences [6].

Repetition time (TR) is the amount of time between successive, nearly identical parts of a pulse sequence.

Time to echo or echo time (TE) is the time between an RF pulse excitation and the location of the echo signal (this is typically the maximal signal produced by a gradient and/or spin echo, for example).

Expounding more on what Dr. Elster described (as mentioned above in section 3.3.1), for clinical imaging clinicians seek to generate a 'weighting' of the image to visualize normal versus disease processes. This is reflected in the type and parameters of MR pulse sequence employed to create a target image that would demonstrate the noticeable differences between pathological and non-pathological tissues. At this stage, we mention TR and TE as two parameters that are modified to create T1, T2, and PD weightings. Below are the general parameters for each sequence, which you may have already heard if you are in your residency or in radiology technologist school:

T1-weighted = short TE, short TR;
T2-weighted = long TE, long TR;
PD-weighted = short TE, long TR.

For further information on this topic, you might continue by visiting this website [4].

Each sequence type aims to highlight a different aspect. Some ways that people describe the general appearance of T1 and T2 on clinical images based on weighting, which we will discuss further in section 3.3.3, follow.

On T1, fluid is dark, fat is bright, and free water tissue is semi-dark.

On T2, fluid is bright, fat is semi-bright, and free water tissue is bright.

There is some imprecision in this description, as there are specialty sequences, such as T2-FLAIR, in which fluid appears dark. However, using weighting to communicate is a way that you can get started with MRI, and is frequently heard at radiologists' workstations.

The following DLS images highlight the differences in parameters used to define sequence types. Focusing on TR for the T1-weighted images (DLS figure 3.11), and TE for T2-weighted images (DLS figure 3.12), what do you see in these images? Keep in mind the anatomy of the brain and varying tissue types. Write down anything that may stand out. We will continue to cover these differences in succeeding sections.

Let's look carefully at figure 3.11 and consider T1 relaxation and the differences between the types (if you have previous experience with brain anatomy):

1. We see that a short TR creates greater contrast for T1-weighting
2. Look at the cerebrospinal fluid (CSF) space and compare that with brain parenchyma. Notice there is a difference between A and B.

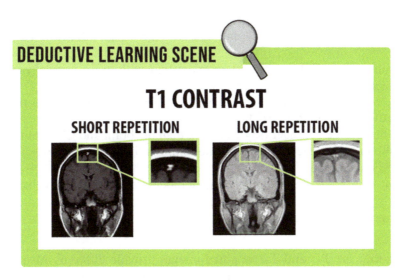

Figure 3.11. Note that on these T1-weighted images, the contrast is high on short TR (A), but low on long TR (B).

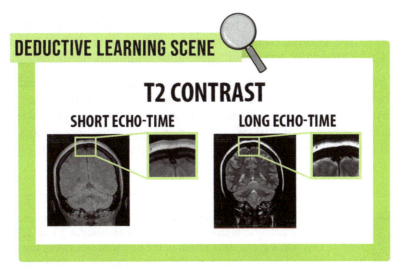

Figure 3.12. Note that on these T2-weighted images, the contrast is low on short TR (A), but high on long TR (B).

3. When the TR (500 ms)† is short, brain and CSF are well separated on T1-weighted images.
4. When the TR is long (2s)†, then the brain and CSF values are nearly the same on T1-weighted images.
5. Fat (the scalp, for example) is bright on both short and long repetition. While not part of this image in particular, workup of fat content can be important to the diagnosis.

Let's next look at figure 3.12 and consider the signal difference based on T2 contrast.
1. Examine the CSF space and compare that with brain parenchyma. Notice there is a difference between A and B.
2. When the TE (10 ms[8]) is short, then brain and CSF are nearly the same on T2-weighted images.
3. When the TE is longer (say 100 ms[8]), then the brain and CSF values are very different. However, on T1-weighted images the contrast between brain and CSF are similar.
4. Fat (the scalp, for example) has a moderately high gray value intensity. While not part of this image in particular, workup of fat content can be important to the diagnosis.

Notice that the T1- and T2-weighted images provide different advantages. By scanning the brain using both sequence types, we can acquire **multiple looks** to gain a greater perspective. Tissues such as fat and fluid as visualized on these images will have unique T1 and T2 parameters that make them pronounced. A key point for technologists and those acquiring the images is that it is important to attempt to complete the entire protocol as best you can. This is because each sequence can contribute to the components necessary to differentiate different tissues and diseases. This is especially important when subtle pathology needs to be diagnosed.

Fat and **fluid** are important components of evaluating an image, so understanding how you can manipulate these (through **pulse sequences**) is important. The appearance in these two tissues first requires some understanding of the basic physics (**relaxation**) that create their inherent contrast. We will go over more of the appearances on images later in this chapter. The next step is to get into the 'guts' of what is going on. For that, we have to discuss longitudinal and transverse magnetization.

Nick P. (third-year medical student) reflects on examples of body parts or conditions that are better to scan using a T1 or T2:
"This was a great explanation of the differences between the different weighted images and how they are obtained. I was always taught that in the brain, for example, T1 imaging is better for picking up masses or structural abnormalities, while T2 is more used for picking up fluid shifts or edema, but it was helpful seeing how the actual physics affects how we obtain the different images."

---
[8] Numbers are provided here because the next section discusses the impact of choosing TE and TR and T1- and T2-weighting.

3.3.4 How does relaxation work? Examining longitudinal and transverse magnetization

If you have mastered the dipole and precession concepts[9], then you have a good start to understanding the appearance (clinically/biologically). Thus, we can introduce a little bit of physics to better understand the behavior. We will describe the atomic/molecular basis of T1 and T2 and then examine their manifestations in the presence of a magnetic field. In brief, T1 is a *regrowth* factor that occurs in the longitudinal direction, while T2 is a *decay* factor in the transverse plane (rotating at resonance frequency perpendicular to the field). A convention is to label longitudinal magnetization as M_z and the transverse magnetization as M_{xy} [7].

Let's start with this picture. Focus on the vector pointing in Z, which will be the longitudinal direction (figure 3.13).

3.3.5 T1 as regrowth (dissecting the details)

For T1, the focus lies in the longitudinal direction. The spins align naturally in the direction of the main field (M_z). When excitation is introduced (from RF pulses), the spins absorb energy and are tipped into the transverse plane. When the RF pulse halts, the spins begin to relax or 'regrow,' back to their natural state. The T1 relaxation is the realignment of the spins with the main field

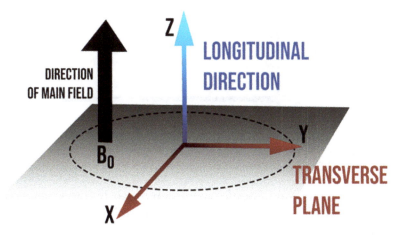

Figure 3.13. Longitudinal direction clarified in this direction. The length of the vector is considered to be the M_z component of the ensemble of spins. Note that spins in the xy plane are rotating with speeds near the Larmor frequency.

---

[9] If you are still having challenges with dipoles and precession, it is possible to understand MRI by examining them qualitatively through relating to the images that we described in section 3.3.2. But, a major benefit of learning the motion of relaxation is that later there are several topics that build upon earlier concepts (i.e., diffusion, time of flight, black blood imaging, etc). A firmer understanding of relaxation and precession will help you better relate and intuit those concepts in more relatable fashion.

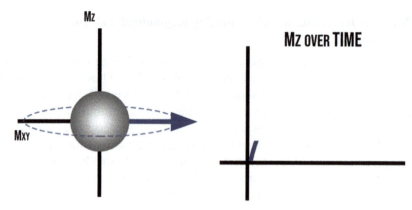

Figure 3.14. After an RF pulse, the proton is momentarily pushed into the transverse plane. The proton then returns to its natural state in the lateral plane, inducing a measurable change in the M_z field. For the purpose of this image, we are observing in rotational frame so it appears to be regrowing without rotating. Animation available at https://doi.org/10.1088/978-0-7503-1284-4.

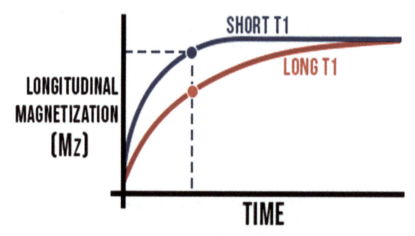

Figure 3.15. Compare two tissues with different T1's, one short and one long. Note that the contrast between the two tissues results from the differences in the T1 values of these tissues.

direction. As relaxation occurs, the energy absorbed by the spins from the RF pulse is dissipated within the lattice (see prior scaffolding bubble in chapter one on the topic of the lattice). This is why T1 is referred to as the spin-lattice relaxation constant.

T1 relaxation is broken down into two vectors. We see an increase of magnetic field within the longitudinal direction (M_z), but a decrease within the transverse plane (M_{xy}) (figure 3.14).

T1 reflects the length of time it takes for the regrowth of M_z to reach its initial maximum value (M_0). Tissues with short T1's recover quicker than tissues with long T1's. Their M_z values are larger, producing a stronger signal and a brighter spot on the MR image. If acquired too early, or too late, there will be little contrast between tissues, as seen in figure 3.15.

Note that field strength influences T1 relaxation times. For example, at 1.5T, gray matter has an average T1 value of 859 ms and white matter of 800 ms. At 3T, the average T1 value of gray matter is 1330 ms and white matter is 830 ms. These values are estimates and vary across institution and location [4].

3.3.6 Scaffolding bubble, visualizing T1

If you are a primarily visual learner and like to have one more way to look at T1, this example provides one more rendition of T1 regrowth. At this point, you should have mastered the concept of the longitudinal direction components and the transverse plane components. Let us watch the action of spin in the animation as it changes in the plane [7].

While the motion of T1 appears to be complex to visualize, if you as the reader are sensing this motion in your mind and in the future are able to relate it to the images, you are connecting more to the 'physics' and meaning behind the acquisition in your 'image' readings [8], video available at https://www.youtube.com/watch?v=lKp67IqQjH4.

Reflection by Gail H. (neuroscience postdoctoral researcher):
"T1 is an important method for obtaining structural information in brain research. For example, neuroscientists often need to distinguish between gray and white matter, which is much easier using T1."

Dr. Wu added:
"I believe there are many uses for T1-weighted scans in the clinic. However, one simple concept is that T1-weighted scans are used for structural images, as opposed to T2-weighted scans, which exhibit features such as edema. This is a vast oversimplification, but a potential way to start to think about relaxation weightings in MRI."

Key points for T1
- T1 is the time constant for regrowth of longitudinal magnetization (M_z).
- Synonyms: spin-lattice relaxation, thermal relaxation, longitudinal relaxation.
- Energy is released from spins to environment ('lattice').
- T1 values vary due to magnetic field strength.

At this point, we have covered the effects of relaxation within the longitudinal direction. The next goal is to consider the relaxation within the transverse plane (T2 and T2*). In the next section, we will consider the 'T2' component known as spin–spin relaxation.

3.3.7 T2 as decay

T2 relaxation can be described as a loss of signal coherence in the transverse plane. This is known as T2 **dephasing**. After an RF pulse, the spins are tipped into the transverse plane (M_{xy}). The motion that we observe in the transverse plane is precessional (rotation at Larmor frequency). The spins are coordinated, same frequency and same speed, resulting in a large net magnetization in the transverse plane.

However, these spins are not naturally coordinated and begin to precess at different frequencies. This spreading out, or dephasing, of spins is the T2 effect of spin–spin interaction. The net magnetization within the transverse plane exponentially decays over time. As time progresses, this sum will get smaller as the vectors begin to spread around the circuit. By adding and subtracting, you eventually will get a decay.

Each spinning proton has its own magnetic field. Neighboring protons directly affect each other. Consider a noisy coffee shop. The person at a neighboring table is speaking loudly on the phone; you hear parts of the conversation. You receive a call; now the neighboring table hears parts of your conversation. The spin of neighboring protons 'interrupts' each other. As this continues, the interactions will cause cancellations, and eventually the overall signal will decay (figures 3.16–3.19).

T2 is the time constant for decay/dephasing of transverse magnetization (M_{xy}). Here again we can evaluate the exponential curve of T2 (exponential decay). Just as before, it is instructive to consider spins with two different field strengths. Note that T2 is not altered by field strength, as T1 is.

T2 reflects the length of time it takes for the MR signal to decay in the transverse plane. A short T2 means that the signal decays very rapidly. So, substances with

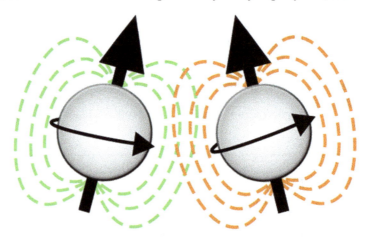

Figure 3.16. Electrical and magnetic fields of individual protons influence the fields of neighboring protons, referred to as spin–spin interactions.

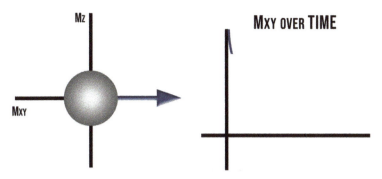

Figure 3.17. Illustration of the effect of the spreading out isochromats. As time evolves, the spin–spin relaxation (T2) will cause the spins to disperse farther and farther apart, such that if you summed the vectors, the resultant sum would continue to diminish in magnitude, as shown in the figure. Animation available at https://doi.org/10.1088/978-0-7503-1284-4.

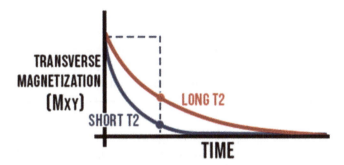

Figure 3.18. Compare two tissues with different T1's, one short and one long. Note that the differences (contrast) between the two tissues result from the differences in the T1 values of these tissues.

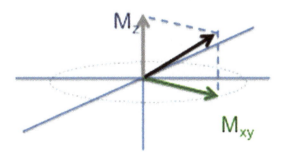

Figure 3.19. The Lab Frame is as if you are standing on the outside and looking at the precession and decay happening at the same time. Animation available at https://doi.org/10.1088/978-0-7503-1284-4.

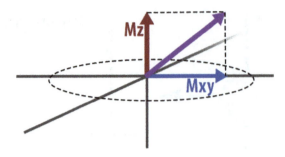

Figure 3.20. The rotating frame is as if you are in the rotating precessional frame (what you will see is only the decay). Animation available at https://doi.org/10.1088/978-0-7503-1284-4.

short T2's have smaller signals and appear darker than do substances with longer T2 values. Again, if acquired too early, or too late, there will be little contrast between tissues, as the graph illustrates in figure 3.18.

Recall that the transverse plane is different from the longitudinal direction, as the former points directly with the main direction of the static field, while precessional motion occurs in the transverse (M_{xy}) plane, as portrayed in figure 3.19. Consider two different vantage points: one where you are looking at the spin component as if you are observing it from the outside (lab frame), and one where you are looking at the spins as if you were travelling with the precession (rotational frame), as portrayed in figure 3.20[10].

Natalie N. (Pre-Med) asked
'Hey Gail, did you get why it's called spin-spin relaxation? I'm having trouble conceptualizing this!'

Gail H. (neuroscience postdoctoral researcher) responded:
"I think it is called spin-spin relaxation because the process involves the interaction between adjacent spinning protons and how they affect each other. So 'spin-spin' refers to two neighboring spinning protons. If you look in figure 3.16, you can see how electromagnetic properties of neighboring protons could influence each other. The signal from each proton gets diminished by interference from its neighbor, hence the 'decay' part."

---

[10] Perhaps consider a merry-go-round. If you were the parent watching the kids on the merry-go-round looking at them turn, this would be the lab frame, versus if you were the kid on the merry-go-round observing what is happening.

> **Scaffolding bubble**
> For those who like to perceive the action more in 'real time': Now that you've seen the longitudinal direction, let's watch the action of spin as it changes in the transverse plane. Refer to this animation of T2 relaxation [9], video available at https://www.youtube.com/watch?v=is8TscwFOvM&list=PL40F1EE0DF59D777A&index=2.
>
>

Some key points for T2
- Synonyms: transverse relaxation, spin–spin relaxation
- T2 is the inherent spin-spin interaction. Neighboring spins in space (magnetic moments) will have an effect on each other.
- The continuing presence of the spins will cause the signal to decay due to loss of coherence. This loss occurs in the transverse direction.
- T2 is not altered by field strength.

If you have reached this point and can somewhat visualize T1 and T2, congratulations! You have added a valuable skill for the intuition of the major relaxation terms. We hope this may serve you well, as relaxation is the main basis of **contrast** in MRI. If you are not able to visualize T1 and T2, it is worth taking a break and then revisiting it at a later time when you are fresh from study, as if you can master it, you will see more in the images you may be interpreting in the future.

3.3.8 When you understand weighting and how it works, where do we go next with relaxation, clinically?

Now with an understanding of the exponential curves for T1 and T2, we can refer back to this figure from the Deductive Learning Sketch (figure 3.11) in section 3.3.2.

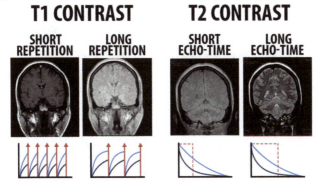

Figure 3.21. Exploring contrast curves and how they affect appearance of T1 and T2. These are the images from figures 3.11 and 3.12, where we demonstrated differences when you manipulate (A) TR on T1-weighted images and (B) TE on T2-weighted images.

When comparing figure 3.21A to figure 3.11 and time, we have now added in the relaxation curves. From these curves, you will note on the T1 images how the short TR (i.e., shorter time elapsed between signals, represented by the red ↑ arrow on the graph) enhances the contrast. Likewise, in figure 3.21B, on the T2 images, notice how the long TE (i.e., larger distance between gradients, represented by the dark and light blue lines on the graph) enhances the contrast[11].

Now that you can connect the curves and the relaxation of spins to how they affect different structures in these figures, we will next proceed to take another step and begin to understand two of the major constituent materials (i.e., fat and fluid) in tissues that alter relaxation on images.

Fat and fluid appearance on MRI

Originally, I wanted to create a table with only non-suppressed sequences, but the ability to manipulate contrast in MRI is extensive, and seeing both suppressed and non-suppressed images seemed to be appropriate here. So, the lower right corner is for the 'advanced' student to become more aware through exposure. But, generally understanding the basic non-suppressed sequence responses to T1 and T2 can be valuable to the clinical reader [10].

	T1	T2
Fluid	Dark (long T1)	Bright (very long T2)
Fluid w/suppression	*Dark, helps contrast gray/white matter in brain*	*Dark, helps contrast with periventricular fluid, which is bright*
Examples in figures 3.22 and 3.23		

	T1	T2
Fat	Bright (short T1)	Intermediate to Bright, depends on pulse sequence, see more info on J coupling effect
Fat suppressed (STIR and/or FatSat Pulses)	*Fat suppressed post-contrast helps with visualizing lesions, such as tumors.*	*Fat suppressed T2 enables better visualization of fluid-containing structures.*
Examples in figures 3.24 and 3.25		

---

[11] If you want to take one step in your understanding of when to get a T1 weighting, to achieve this weighting you would also want a short TE, which would actually diminish the capacity of the T2 effect. Alternatively, to provide a good T2-weighting, you would want to implement a long TR, which would diminish the T1 effect, so you would get a better image.

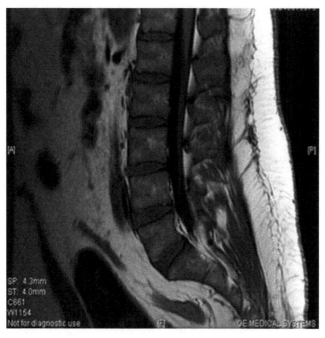

Figure 3.22. T1 FLAIR on spine. Improves clarity of abnormal tissue in bone marrow and spinal cord. Better contrast at the CSF–cord interface and the nucleus pulposus (inner annulus fibrosis)–outer annulus fibrosis interface.

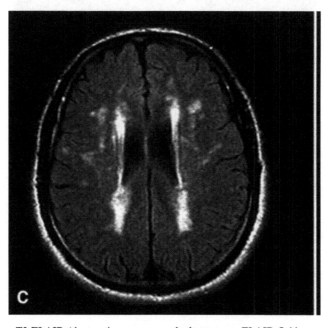

Figure 3.23. MS on T2-FLAIR (the two images are on the bottom are FLAIR-fluid suppressed, and the top two images are regular FSE). On FLAIR, the T2 suppression aids in visualization of periventricular lesions, such as seen on this image.

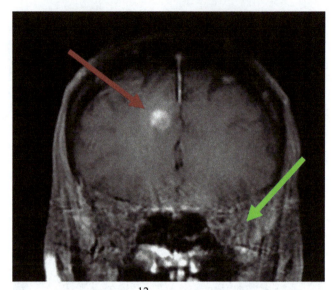

Figure 3.24. T1 Brain with fat sat post-contrast [12], the fat in the brain is suppressed. Fat sat on post contrast T1 helps you visualize neovascularization better (red) without the interference of fat. Note that the fat signal from the orbits is suppressed (green).

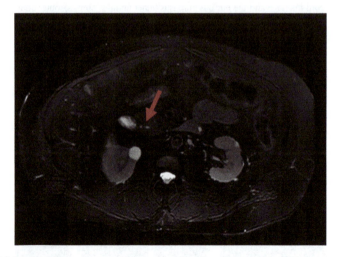

Figure 3.25. T2 fat sat axial sequence used in pancreas imaging, for example. This sequence enables better views of the structures, including the head and pancreatic duct.

---

[12] Fat suppression can be used to suppress the signal from adipose/fat tissue. Fat saturation can be used to suppress fat to help create improved visualization of contrast material-enhanced [11].

Scaffolding bubble

Radiologists digging into shades of tissues

Eventually, after many years of looking at a number of medical images, you can develop a very sophisticated understanding of tissue contrasts (shades). A sample reproduced from the publication 'MR pulse sequences: what every radiologist wants to know but is afraid to ask' [10] can represent the level of complex understanding that a radiologist obtains after years of practice. I'm always amazed that attendings manage the numerous complexities of tissues, for each subspecialty has their challenges.

Recognize that different tissue content enables different forms of weighting. This could include mineral-rich, free water, bound water, lipids, proteinaceous tissue, blood, and more.

Casting materials, tissue physiology intermixed with the complexity of disease, and being able to recognize those patterns is quite a lifetime accomplishment.

Examples levels of T1-weighting include: dark: air, mineral-containing materials, and/or fast-flowing blood; low gray: collagenous tissue, bone islands, tissue with non-bound/free water (edema, simple cysts, bladder, gallbladder, CSF), bound water tissues (such as some organs and muscle); intermediate gray: protein-rich constituents (some abscesses, complex cysts, synovium); and bright: fat, fatty bone marrow, methemoglobin in blood, melanin, paramagnetic contrast agents. Example levels of T2-weighting in tissues include: dark: air, mineral-rich tissue, fast-flowing blood; low gray: collagenous tissue (ligaments, tendons, scars), boney islands, high bound water tissues (for example, the liver, pancreas, adrenals and some cartilage and muscle), fat, fatty bone marrow; and bright: non-bound/free water tissue, protein-rich constituents, blood products (oxyhemoglobin and extracellular metHb) [5].

One major reason it is important for radiologists to understand these 'shades of tissue' on images is that they may be asked by a referring physician (non-radiologist physician, such as neurosurgeon, oncologist, and/or orthopedist) to help them diagnose underlying pathophysiological processes. For this reason, translating or turning the 'gray' scale assessment to something meaningful in terms of constituent matter (fat, fluid, protein levels, etc) can help guide the referring physicians to better understand the pathophysiology and makeup of the body that is being imaged.

Justin reflects on the appearances in MRI

"When I am looking at an MR I am constantly asking myself, 'What is the image telling me.' Bright and dark aren't just bright and dark to me. Tendons tend to be very dark, but tendons with calcification are extra dark. Fat is bright on T1, but the fat in a lipoma will be subtly brighter due to higher fat content in a lipoma versus normal adipose. I really rely on subtle differences to make the diagnosis. One of my favorite things is watching residents learn what subtle edema looks like in muscle on a STIR image. It is very easy to miss, but once you have seen it a few times, it becomes much more obvious."

As you advance your knowledge in the field, you may eventually become more aware of different tissues with different constitutive makeup, including mineralized, proteinaceous, free and bound water tissues, and more.

Applied bubble
A PD image applies a medium TE, while having a Long TR. While less frequently used these days, PD images are often utilized in the area of musculoskeletal (MSK) imaging for the evaluation of bones and joints (figure 3.26).

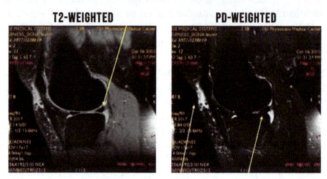

Figure 3.26. MRI scan of 'housemaid's knee.' The PD weighted image provides further evidence of bursitis and degradation of cartilage [6].

PD is also sometimes used for neonates, when myelin changes are less pronounced, to obtain more contrast in the image. In some cases, PD-weighted images in neonates can be useful in the evaluation of gliosis and/or hypomyelinating conditions. For more reading on PD imaging, please see the appendix.

Scaffolding bubble:

Nick P. asked, "What is a PD Scan?"

Dr. Wu replies with Dr. North:
"PD-weighted scans are not used as frequently, so that's why we focus on T1, T2, and T2 star imaging in this chapter. PD scans look at the number of spins available (the sheer number of hydrogen protons), as opposed to magnetic characteristics of the protons (like mechanisms of

relaxation) that are used in T1 and T2 weighted images. The main applications are in MSK, where subtleties between structures such as cartilage and bone are better visualized. Tissues that have high proton density include fats, while different types of cartilage have medium to low densities, so the scans are useful in delineating these tissues. PD scans are often also used in pediatrics, especially neonates, because they are able to show discrimination between tissues where myelination hasn't fully formed. PD-weighted images permit some of the scans to have more SNR, however, because they have decreased TE."

Summary for relaxation

MRI signal is largely influenced by several forms of relaxation (T1/T2) to create the inherent contrast of the image. The ability to differentiate tissues and processes is a key part of diagnostic radiology. Therefore, relaxation should be one of the first concepts for understanding clinical MRI. Many radiologists spend many years studying the appearance of disease. A mix of perspectives between the acquisition physics and the clinical manifestation can be helpful.

A simplified understanding of relaxation could be that the T1 signal is the return and regrowth to original parallel/antiparallel position. The energy gets transferred to the lattice (i.e., the surrounding environment as a whole). T2 is the cross talk between neighboring spins. Remember that changing the M field changes the E field, and vice versa. Two spins at different positions will have slightly different offset frequencies. As they twirl together, slightly off in phase from each other, they will affect each other. In the presence of each other, they will eventually create decay.

The following summary table can take this basic understanding one step further:

T1 appearance on clinical images	T2 appearance on clinical images
T1-weighted imaging is for visualizing normal anatomy. *TR:* Short *TE:* Short *Fat:* Bright *Fluid:* Dark	T2-weighted imaging is for visualizing fluid/edema. *TR:* Long *TE:* Medium *Fat:* Bright/Gray *Fluid:* Bright
T1 relaxation: • Spin-lattice or longitudinal relaxation • Typically a few hundred milliseconds • At time equal to T1, 63% of longitudinal signal has returned	**T2 relaxation:** • Spin-spin relaxation or transverse relaxation • Typically tens of milliseconds • At time equal to T2, decays to 37% of maximum

(Continued)

(Continued)

T1 appearance on clinical images	T2 appearance on clinical images
• T1 increases approximately linear with field strength • Energy transferred from spins to environment (lattice) permitting regrowth • Contrast agents, such as Gadolinium-DTPA, cause T1 shortening (brighter)	• T2 little dependence on field strength • Signal coherence is lost through spins interacting with other spins

Yield bubble

3.3.9 Application of imaging weighting (given TE, TR parameters)

Now that you have mastered what relaxation is, it's time to practice being able to identify the intended weighting of the sequence based on TE and TR parameters. Images are sometimes mislabeled because of automation of the scanner and not caught by the technologist, and are then put in the wrong category. It is important to recognize this as it would be like an 'artifact' in terms of consequence. Therefore, this topic is often tested on board exams (table 3.1).

Table 3.1. A table of various sequence parameters. Note TI (inversion time) is discussed in other references.

Parameter	Answer
TE = 8 TR = 600	T1 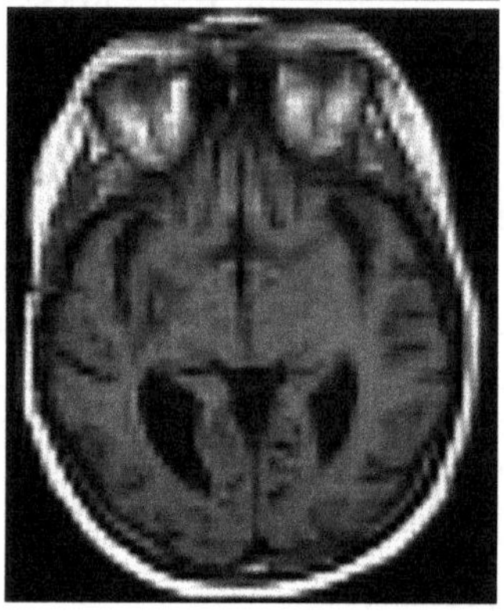

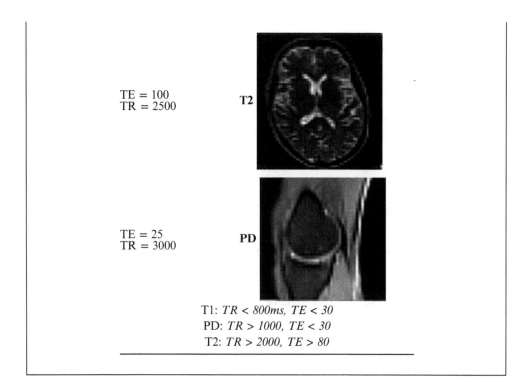

T1: *TR < 800ms, TE < 30*
PD: *TR > 1000, TE < 30*
T2: *TR > 2000, TE > 80*

Knowledge check (assume a 1.5T scanner):
1. Explain how changing the value of TE from 15 msec to 80 msec affects image contrast.
2. Explain the general effect of changing TR from a value of 300 msec to 1000 msec on the signal intensity and relative brightness of an image.

Suggested answers
Question 1 is slightly a trick question. In order to answer the question, you would want both the TR and the TE values. A TE of 15 would yield a T1-weighted image. Changing the TE to 80 would yield a T2-weighted image. However, it is important to consider both parameters because under the same TE parameter of 15 msec, a TR of 300 would likely yield a T1-weighted image, while a TR of 2000 msec would yield a PD-weighted image.

In Question 2, changing the TR value from 300 to 1000 msec would increase the signal intensity and brightness on the image. However, if the TR of the image was 1000 msec and the TE of the image was 20 msec, it would generate a PD-weighted image, whereas a TE of 100 msec would generate a T2-weighted image.

The author intentionally left the details out, because the author is hoping that the reader develops an instinct to question parameters and is able to rely on their concepts to carry them further in MRI.

Nick P. reflects on how would you go about memorizing these factors so you can identify the T1 and T2 sequences:
"When examining the parameters to determine the weighting of the image, it is important to use both the TE and TR values. I found it helpful to think of T1 imaging with lower TE and TR values, and T2 having higher values. PD seems to be a low TE with higher or middle-range TR values."

3.4 Including inhomogeneity T2* (into T2)!

Learning objectives:
1. What it looks like (DLS), T2* as producing undesirable artifact, T2* as a diagnostic tool.
2. What is meant by susceptibility?
3. T2 inherent versus Inhomogeneity term - spin/spin.
4. How the inhomogeneity term contributes (speed of the gradient dephase).
5. T2*<T2<T1 facts to remember.

3.4.1 What does T2* effect look like on images?

What do you see in the DLS? On the left, you see two effects from susceptibility (figure 3.27).

First, some of the images appear to be distorted geometrically. Second, you may recognize the signal loss in the images. Note that geometric distortion and signal loss will be revisited in the spin echo portion of chapter 4 and the geometric distortion portion of chapter 7.

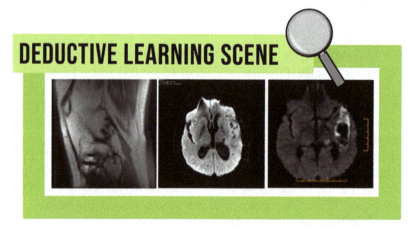

Figure 3.27. A DLS for understanding T2*. Please look over the figures and consider what you see in the image.

Geometric Distortion with signal pileup at interface

 Signal Loss (due to metal implant)

However, there are also occasions where the signal distortions and loss become helpful. For example, this is true in the case of blood products, such as hemosiderin. T2* weighted sequences, such as susceptibility weighted imaging and gradient echo methods, can be affected by the presence of hemosiderin. Note that T2* weighted sequences are commonly used to exacerbate the sensitivity to susceptibility effects. Refer to this article for a helpful mnemonic concerning aging blood on MRI [12].

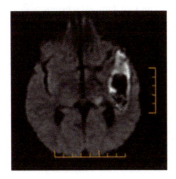

3.4.2 What is meant by susceptibility?

What is a local inhomogeneity? The path of most waves is dependent on the medium they are traveling through, including light. This is due to different materials having different refractive indexes, causing light to pass through them at varying rates, thus creating a distorted image. The same phenomenon exists in MRI[13], and is referred to as **susceptibility**. Susceptibility is a measure of how much a material will be affected (or magnetized) by an applied magnetic field. Different materials (metal implants, cavity fillings, air-filled cavities) have different magnetic susceptibilities compared with surrounding tissue, as illustrated in figure 3.28. This causes the magnetic field to bend and become inhomogeneous. The bending causes artifacts, such as distortion or signal pile up and loss. The bending can be modeled as an additional gradient, as depicted in figure 3.29.

You may recall the exponential decay of waves in section 1.1.6 of chapter 1. We also previously demonstrated the effect of decay on T2 curves [14].

---

[13] Light is an EM radiation. For those interested in 'the why,' it can be instructive to learn about the mechanism of bending waves as they pass through different materials in the following video 'Why does light bend when it enters glass?' [13].

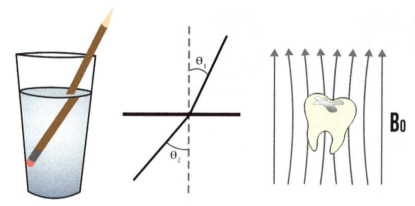

Figure 3.28. Light bends when meeting a material with a different refractive index [8]. Similarly, magnetic field lines bend when passing through a material with different magnetic properties, such as a metallic cavity filling.

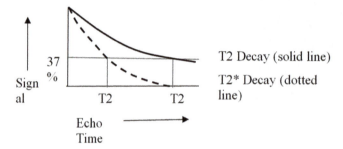

Figure 3.29. Note that the T2* decay induces a faster decay rate than T2 signal. This is primarily due to $\gamma \Delta B$, or the presence of inhomogeneities.

 Gail H. (neuroscience postdoctoral researcher) reflection on field inhomogeneity effect:
"I think it makes a lot of sense that different types of tissues could change signal that is passing through. I remember talking about light refraction in physics class, and it is similar. If you have light passing through air and then through water, there is a bending of light. So, if you have electromagnetic signal going through soft tissue, then bone or air, it could similarly bend the signal and cause a distortion. Very interesting!"

3.4.3 T2 versus T2*

At this point, it's helpful to provide the equations for T2* (figure 3.30). This can be expressed as:

$$1/T2^* = 1/T2 + \gamma \Delta B \tag{13.4}$$

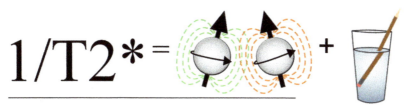

Figure 3.30. A visual expression of the intrinsic spin-spin decay with the additive component of local inhomogeneity.

T2* is equivalent to T2 plus the consideration of inhomogeneity. When the magnetic field is entirely homogeneous, 1/T2 = 1/T2*

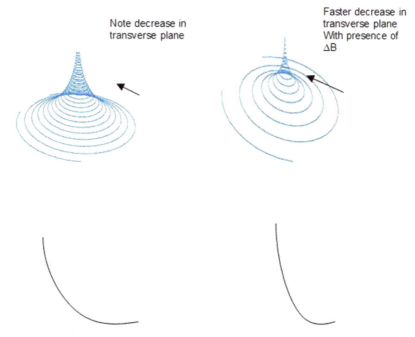

Figure 3.31. Faster spin down with T2*. Shown is T2 combined with a T1 effect so that one can visualize the effect of the transverse spins in combination with T1 for potential improved understanding of the combination of effects of relaxation.

Recall that T2 was defined as a time constant for the decay of transverse magnetization, arising from natural atomic or molecular interactions (figure 3.31). This is represented by the spin–spin component on MR. In any real application, the transverse magnetization decays much faster than would be predicted by natural atomic and molecular mechanisms; this rate is denoted **T2***. This

Figure 3.32. T2* Animation that has a similar decay pattern to that of intrinsic T2 decay, but a faster decay rate [14].

relaxation term results principally from inhomogeneities in the main magnetic field, which includes both the inhomogeneity effects and the natural T2 spin–spin interactions. Figure 3.31 illustrates a comparison between natural decay (no ΔB component) and the faster decay with presence of ΔB. These inhomogeneities may be the result of intrinsic defects in the magnet itself, from susceptibility-induced field distortions produced by the tissue, or other materials placed within the field that can be part of an inhomogeneous environment.

3.4.4 How does inhomogeneity term contribute?

T2* considers the dispersion of spins as a result of the addition of local field inhomogeneities that creates faster dispersal when compared with T2 (i.e., intrinsic spin–spin decay), i.e., $1/T2^* = 1/T2 + \gamma \Delta B$. Note that increasing the inhomogeneity (ΔB) will decrease the T2* value. An advanced treatment of this is seen in chapter 20: 'Magnetic field inhomogeneity effects and T2* dephasing' of Brown *et al* [15] (figure 3.32).

> **Applied bubble**
> Here is where the reader might consider connecting their knowledge of the main magnet and shimming from chapter 2. Often, the best location for scanning is where you would have the greatest homogeneity. At the center of the bore, you would have best signal in the middle of the main field bore, and often the case for exams is to carefully locate the patient's wrist or elbow in the center of the bore to achieve the best signal free from distortion and signal loss (due to the inhomogeneity effects we just discussed).

3.4.5 Summary

1) Note that T2* is a combination of T2 and local homogeneity (ΔB).
2) Note that as ΔB goes up, T2* goes down. This would mean the decay rate goes down.
3) Can you think of the effect of homogeneity (increasing it changes the rate constant, making it shorter)?
4) T2* relaxation is much shorter than T2 (a few milliseconds in scale).

5) The T2 component of relaxation is irreversible (lost with time).
6) If the magnetic field inhomogeneity term is pronounced, it causes (T2*) blooming susceptibility.
7) T2*<T2<T1

3.5 Table of T1/T2 and physical values

A while ago, it was not uncommon to have concise values describing structures in the body. Nowadays, most radiologists need just a relative sense of the values (not necessarily exact numbers). For example, a relative idea of values can be understood as fat being around 4× shorter than gray matter at 1.5 T. However, due to advances in technology and computing power, there are a few applications where T1, T2, and T2* quantitative measurements can be used in the diagnostic process. In these applications, a table of values based on peer-reviewed literature with ranges on which one can base clinical recommendations can be used. Thus, quantitative MRI generally stays within the academic sector, but it is continually improving as MRI manufacturers have placed the applications into FDA-approved packages.

From Allen Elster (see MRI.questions.com), which we also believe is good guidance:
'I expect learners to make at least three observations: (1) that for most organs T1 values are typically about 5–10× longer than T2 values; (2) that pure liquids (water/CSF) have very long T1 and T2 values; and (3) that dense solids (ice, tendons, and proteins) have very short T2 values.'

Tissue	T1 (ms)	T2(ms)
Gray matter	950	100
White matter	600	80
Cerebrospinal fluid (CSF)	4500	2200
Muscle	900	50
Fat	250	60
Blood	1200	150 (can vary)

REMEMBER: these tables are useful in estimating the signal for certain sequences. Mapping is just one way to get the values. Perhaps more useful is to be able to visualize the decay rate and regrowth rate scales to reinforce your understanding of when the signal is bright and dark. Be aware that the values differ between sources, so this is only intended to get a general understanding of relative values.

Literature bubble
In the early days of MRI, there was an attempt to broadly estimate relaxation parameters for various organs. Until recently, the practice had grown somewhat out of favor. However, with the advent of new techniques and improved computational processing, there is new interest in quantitative applications.

> There are now parametric maps[14] that use the measurements. One map is used in cardiac MRI and is good for evaluating cardiac amyloidosis (T1 mapping). Another is used for the liver and is used to evaluate hemochromatosis (T2* mapping). MRI manufacturers have also implemented T2 mapping sequence and processing utilities that can be used to non-invasively estimate variation in the collagen component of the extracellular cartilage matrix.

3.6 Summary

This was a very dense chapter, but it contains the building blocks for what you need to know later. There are two ways to potentially organize your thoughts around the chapter. One is from the 'ground up,' understanding contrast in images and their values and appearance. The second is the way that the image weightings come about through the selection of TE and TR. If you have a 'starting' handle on concepts such as contrast, resolution, T1/T2/T2* relaxation, and weighting, then you are beginning to understand a fundamental part of MRI.

Holding the information in your head may permit you to understand the deeper parts of this book, including making tradeoffs/optimizing sequences, going faster, recognizing artifacts, looking at physiology in more detail, and getting a glimpse of some of the cutting-edge clinical applications and research. *Contrast, spatial resolution, relaxation. Can you partly visualize the motions of spins?* M_z points up with the field, which is the longitudinal direction. M_{xy} is in the plane that is perpendicular to M_z, and is called the transverse plane. Now that you have an understanding of the longitudinal direction and transverse plane, we can proceed into further descriptions of how they work.

Next, we practiced with real-world parameters by looking at a list of factors TE and TR to set up parameters. We then discussed the incorporation of inhomogeneity terms into the T2 so that we have T2*.

Finally, we included some practice with general numbers for relaxation values. The important part may be to just have a relative sense of the values (memorization of the values). There are a few quantitative applications where estimates for T1, T2, and T2* may lead to a diagnostic criterion. We mentioned that it would be informative to be open to those few applications, such as cartilage mapping, amyloidosis, and iron overload mapping.

References

[1] Step-by-step building guide (n.d.) Thehousedesigners.com (retrieved 26 January 2022) https://www.thehousedesigners.com/articles/homebuildingguide.asp

[2] Resolve definition and meaning—Merriam-Webster (n.d.) (Retrieved 26 January 2022) https://www.merriam-webster.com/dictionary/resolve

---

[14] Parametric mapping is the process of assigning a value at each point of the image that a parameter represents the quantitative value estimated. Relaxation values (e.g., T1, T2, or T2*) are the types of maps that are described as the parameter of interest in this section.

[3] Burathoki S 2014 Understanding brain aneurysm and endovascular coiling https://youtu.be/ahCt7hNT4Zc?t=256
[4] Image contrast (n.d.) Questions and Answers in MRI (Retrieved 26 January 2022) http://mriquestions.com/image-contrast-trte.html
[5] Tanenbaum L N *et al* 2017 Field synthetic MRI for clinical neuroimaging: results of the magnetic resonance image compilation (MAGiC) prospective, multicenter, multireader trial *Am. J. Neuroradiol.* **38** 1103–10
[6] Parameter 'weighting' What is meant by a T1- or T2-weighted image? (Retrieved January 26, 2022) https://www.mriquestions.com/meaning-of-weighting.html
[7] Drew Z (n.d.) Longitudinal and transverse magnetization (Radiology Reference Article, Radiopaedia.org) (Retrieved 26 January 2022) https://radiopaedia.org/articles/longitudinal-and-transverse-magnetisation?lang=us
[8] kas pijpers 2010 MRI principles 1 of 4–90 degree pulse and T1 relaxation.avi. https://www.youtube.com/watch?v=lKp67IqQjH4
[9] kas pijpers, 2010 MRI principles 2 of 4—T2 dephasing.avi https://www.youtube.com/watch?v=is8TscwFOvM&list=PL40F1EE0DF59D777A&index=2
[10] Bitar R *et al* 2006 MR pulse sequences: what every radiologist wants to know but is afraid to ask *RadioGraphics* **26** 513–37
[11] Delfaut E M, Beltran J, Johnson G, Rousseau J, Marchandis X and Cotten A 1999 Fat suppression in MR imaging: techniques and pitfalls *RadioGraphics* **19** 373–82
[12] Sheikh Z (n.d.) Aging blood on MRI (mnemonic) (Radiology Reference Article, Radiopaedia.org (Retrieved 26 January 2022) https://radiopaedia.org/articles/ageing-blood-on-mri-mnemonic?lang=us
[13] Fermilab 2019 Why does light bend when it enters glass? https://www.youtube.com/watch?v=NLmpNM0sgYk
[14] Moderate Republican 2013 T2* MRI :-) https://www.youtube.com/watch?v=wHLje7mc8RY
[15] Brown R W, Cheng Y-C N, Haacke E M, Thompson M R and Venkatesan R 2014 *Magnetic Resonance Imaging: Physical Principles and Sequence Design* 2nd edn (New York: Wiley)

IOP Publishing

MRI: Connecting the Dots
A start to concepts
Dee Wu

Chapter 4

The inside details of MRI

Keywords: free induction decay (FID), pulse sequence diagram, spin echo

4.1 Introduction

Building on our original construction analogy, we have constructed the 'exterior façade' of the house and it is now time to move to 'interior' of the house. The concepts discussed in chapter 3, contrast and resolution, relaxation, T1 and T2 parameters, and T2*, will provide necessary insight to the upcoming topics in chapter 4. We will now proceed to discuss three concepts in this chapter: **free induction decay**[1], pulse sequence diagrams, and the spin echo pulse sequence, as portrayed in figure 4.1. Pulse sequence diagrams (PSD) are figures that illustrate the sequence of events that occur during magnetic resonance imaging. PSD provides a timing diagram that illustrates when the timings of the radio frequency (RF) pulses gradients waveforms (slice, frequency, and phase) and the acquisition window are implemented. The PSD structure is essential to determining the final image weighting and contrast in MRI.

When working toward the ideas included in this chapter, it is important for the reader to better understand that there are two major forms of signals in MRI. Explicitly, the signal from MRI has these two methods: (1) *free induction decay (FID)* and (2) **echo** *(spin echo is most common, but a stimulated echo is also considered to have the echo form)*.[1] In this section, we focus on the **free induction decay** (FID) to begin to better understand MRI.

---

[1] There are a few different signal pathways to generate an image from the RF (radiofrequency) excitation. The signal could come straight off the tipped RF excitation, which is the FID described. Another signal pathway could be from an 'echo' signal, which comes from a combination or set of at least two RF pulses. The FID is one of the simplest signals to generate, so it appears first in this chapter. Later on, in section 4.4 of this chapter, we will describe the spin echo.

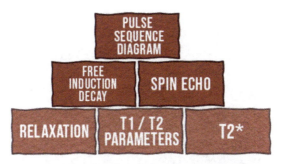

Figure 4.1. Building blocks for continuing to learn MRI, as shown in the bottom row. This illustrates that FID, pulse sequences, and the spin echo will be built upon concepts of chapter 3, which included relaxation, T1/T2 practice with realistic numbers, and the concept of T2*.

4.2 Free induction decay (FID)—your first signal

The FID is the first concept you will approach that demonstrates an output before you engage with more complex tasks. The main concept of this section is that *the FID represents the signal received directly after excitation from the RF (radio-frequency) pulse.*

> Another way to describe the FID is how Radiopedia[2] describes it:
> "... a short-lived sinusoidal electromagnetic signal which appears immediately following the 90° pulse."

Consider this FID to be analogous to the **'Hello world'** program [1] shown in figure 4.2. The understanding of the FID is the initial gateway to your MRI understanding. Other examples of gateway concepts from different domains of knowledge include when you may have written, the first dish you cooked without assistance, your first piece of commissioned art, the first MRI image you read of a patient, or the first patient you scanned on with MRI.

It is useful to describe the free induction decay (FID) first for several reasons. While there are multiple pulse sequence designs, the FID is the most simple and basic. It also demonstrates the T2* decay that you get off a single RF pulse, as well as the wavelike structure that we measure in MRI. To assist you with learning the basis of the simplest part of a pulse sequence, consider reviewing and understanding the 'recipe' to create the FID in the following section 4.2.1.

---

[2] Since 2005, Radiopaedia has grown and is a well-known radiology reference website that has as its mission to create the best radiology reference the world has ever seen and to make it available for free, forever, for all.

Figure 4.2. A Hello World[3] program, as shown by users. Traditionally, it refers to a very simple trial that is used first to test things out [2]. This 'HelloWorld Maktivism ComputerProgramming LEDs' image has been obtained by the author from the Wikimedia website where it was made available by User:Glogger under a CC BY-SA 4.0 licence. It is included within this article on that basis. It is attributed to UserGlogger.

To improve the understanding and communication of this topic, it is useful to look at how other healthcare resources describe the Free Induction Decay (FID).

Gail H. (neuroscience postdoctoral researcher) asked "Can you restate what is an FID and why it is important?"

Dr. Wu said:
"FID is important because it is the first signal that we observe. Coming from the initial excitation, it demonstrates the T2/T2* relaxation decay. Ultimately, it provides the reader with a sense of what is happening during the period right after the RF excitation (*Gradient Echo (GRE)* n.d.,)"

After some work to visualize and understand the concepts of waves (chapter 1) and decay from relaxation (chapter 3), which is made possible by the hardware (chapter 2), you will begin to better intuit these concepts and apply them to your clinical understanding.

---

[3] The 'hello world' experiment is taken from programming computer analogy. The tradition of using the phrase 'Hello, World!' as a test message was influenced by an example program in the seminal 1978 book by Brian Kernigan [2].

4.2.1 A simplistic recipe for producing an FID

Incorporating the knowledge from our previous chapters on core concepts and hardware, we can understand a more sophisticated 'recipe.' Ingredients include Main Magnetic Field, Transmit RF, T2* decay, Readout, exponential curves, and damped oscillation[4].

1. The main magnetic field creates a uniform field (B_0).
2. Spins align along the main field.
3. Set-up 'receiver coil' to listen to signals in the transverse plane.
4. Tip the spins through excitation using the transmit RF pulse.
5. The spins will precess, generating their own magnetic field. As time carries on, the spins will begin to decay and align along the main magnetic field once again. Use the receiver coil to collect the magnetic signature.
6. The magnetic field will induce a current within the loop (i.e., magnetic flux).
7. This current in the loop produces the signal of interest, the simplest response will be the FID[5].

The FID can be visualized, like as shown in figure 4.3.

Can you imagine the inventors and their excitement when they saw this RF-induced signal for the first time? This enthusiasm is still present in even the most senior of practitioners when they observe the raw FID signal. It is the first task resulting in an observable result and can be a sanity check for service engineers when checking out the operations of the scanner. Let us continue to follow in the steps of

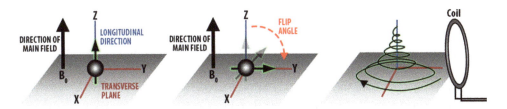

Figure 4.3. To visualize the steps of FID, assume we start in the longitudinal direction. It starts with the tips pointing in B0 direction. The 90 degrees tip it into the transverse plane. Over time, the T2* relaxation decays that signal. The coil sees the oscillating and decaying response, which is converted into the signal known as the FID.

---

[4] Consider reviewing waves concepts section 1.1 in chapter 1 if these concepts seem a little unfamiliar at this point.
[5] After you feel somewhat comfortable with this knowledge, please review the concepts in the section 4.2.2 for a tabular summary of two key points on the FID.

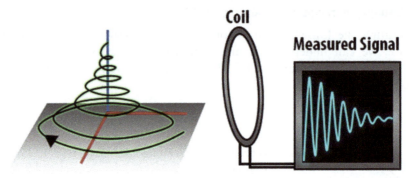

Figure 4.4. Your first signal/spin physics (physics is the study of motion), and the above is what an FID may look like. Notice this is a very similar shape to the exponential decay of a sinusoidal response. If you are curious to see an FID acquired on a small MRI machine, please watch Dr Callaghan's video, where he acquires a live FID pulse [3]. Note the decaying oscillatory pattern (shown in blue) in the left is the pattern of signal that is produced by a **Free Induction Decay.**

the initial MRI inventors and go into what this signal may look like in real life, as portrayed in figure 4.4.

4.2.2 Summary: two key memorization points for FID

1. Remember for the FID you are listening to the 'decaying' oscillations right after the 90-degree excitation pulse. Also recognize the shape of the free induction decay signal.

   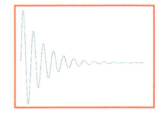

2. Note the elements needed for FID. First, you need the MR Hardware[6] needed to create a signal, such as the **Main field**. Focus on the **RF transmitter** that provides the tipping and the coil that is part of the **RF receiver** system.

   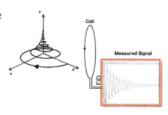

---

[6] Please consider reviewing hardware concepts in chapter 2 that are part of the MRI environment. The reader may also find it useful to review the concepts of Faraday's law ($\Delta M \rightarrow E$) and Ampere's law ($\Delta E \rightarrow M$) that were briefly introduced in section 1.4.

One element to build your MRI learning skills is the key and **central learning idea** that revolves around the understanding that

---

It is inherently possible to <u>encode information of an object through frequencies. The latter to retrieve the information from a collection of waves.</u>

Or more simply put in another way: <u>you can turn information into waves and then there are ways to decode that signal</u>

Natalie N. (Biomedical Engineer and Pre-med) asked:
"Why is the statement above emphasized? What makes it important?"

Dr. Wu answered:
"It encompasses a central but abstract idea that provides key intuition for the readership. Information can be encoded through a range of frequencies, such as through a bandwidth that carries multiple waves over a range. Based on the amount of signal present, all of these waves and frequencies are scrambled together with different amplitudes and transmitted over a distance. This transmission comes in the form of a magnetic field, which carries the waves. By changing the magnetic field to an electric field, we can receive the scrambled signals and put them into a mathematical machine. We can then descramble this signal through an inverse Fourier transformation and retrieve the initial encoded signals."

If you have a basic intuition on the ideas surrounding waves, it will help you **connect the dots** on many concepts in this chapter.

4.2.3 Scaffolding Bubble: motivations for waves revisited

Our readers don't have to come with high aptitudes in mathematics and/or mastery in high levels of physics to grasp some important ideas. By learning concepts, the reader will better understand MRI. This is especially true if they work toward gaining intuition of ideas and do not rely on memorization alone. We also believe that *to demonstrate success at this stage of your journey* is to begin to express and 'feel' and/or 'sense' that mapping a space of frequencies in MRI provides the foundations that create the images we use for the clinical information from our patients.

If you started reading in chapter 1, or perhaps now you are willing to go back to chapter 1 and review wave concepts, it would be a great time to reflect on the things you have learned so far. Perhaps with a little encouragement, it is a great time to reread or refresh your knowledge of waves. By understanding foundational ideas regarding frequencies, phase, and waves, we believe certain pieces of MRI will start to come together. But now, we must turn these ideas of waves into practice! With skills in the understanding of waves, maybe the reader can envision greater meaning in images beyond just pure gray scales. Note these waves are manipulated by the pulse sequence diagrams to produce different weightings/contrasts through the timing of the **gradients and RF (radiofrequency)**. Finally, if you can take the next step to integrate physiology, function, metabolism, and the medical underpinnings with basic knowledge of how MRI works, you will construct understandings that yield a more 'connected', evidence-based, efficient process that will then lead to a more informed approach that we believe will continue to assist your career.

Let's restate some facts to consolidate this logic on Fourier Transform and waves[7].

There are at least two reasons why waves are so important to MRI from a clinical point of view and that pulse sequence diagrams provide the timing that is required to create the image contrast:

(1) Waves are the basis of how we encode in MRI. We manipulate waves to achieve **MRI Tradeoffs**[8] in scanning and are highly dependent on the structure of the pulse sequence diagram.

(2) You also need to understand waves as they are highly relevant to setting the **MRI image weighting** (i.e., contrast in images), which is highly dependent on waves through the parameters we select, as well as *relaxation*[9].

It is natural that you may require more time to study and understand this information and let it percolate. Wave concepts can feel a little out of the ordinary course of your daily lives. As we mentioned in chapter 1, while bats and dolphins feel very comfortable with waves, we may not instantly see the patterns[10]. As humans we must work out these concepts, as we will initially feel uncomfortable with the intuition of waves. However, with a little more work you will grow and better understand these concepts that are particularly relevant to pulse sequence design. This makes this subject a very important and fundamental topic that can help the reader develop a 'deeper' connection in their MRI understanding. Give yourself a break, but keep trying and discuss the topic with others as you go through your journey in MRI.

---

[7] For simplicity, the author has recommended that you use this idea to help you understand the basic concepts of the gradient in pulse sequences. It does have other roles, such as traversing k-space, and even roles beyond the scope of this book, such as spoiling and refocusing. But, we hope to keep things simple and reserve the possibility to describe more detail on the gradient so that earlier learners can maintain the base concepts before we continue to the more nuanced ideas.

[8] Tradeoffs are a highly board testable and clinically relevant topic, and covered in chapter 6. This chapter includes improving signal-to-noise and making the total time of a time of the scan more efficient.

[9] Relaxation is covered in chapter 3.

[10] Please see one that was recommended by one of the talented OU Health diagnostic radiology residents who recommended the 3BlueBrown video on Intuitions on the Fourier Transform and Inverse Fourier Transform [4] and for interactive practice, the PHET website (as we showed previously). We tell our residents that it is normal not to immediately 'see' this connection, but if you were a bat and/or dolphin, it would become more comfortable.

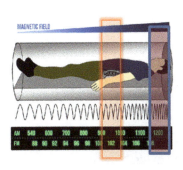
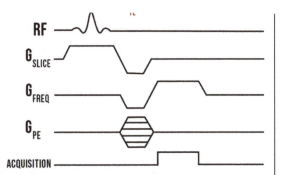

Figure 4.5. Illustrates selecting a location in the body using a gradient. Note that in this picture we are showing the bandwidth of frequencies that are mapped to a field of view or length of the region scanned. On the right, there is a picture of the pulse sequence design, which is the next topic to learn as part of the bricks in the house of your understanding of MRI [5].

Let's look at the following figure integrated with the pulse sequence diagram, as shown on the left of figure 4.5. This section will reveal some of the fundamentals you need to know to better understand MRI.

Next, we plan to provide the briefest of introductions to **pulse sequence diagrams**. If you understand the basis of pulse sequences, it will be easier to understand how parameters are manipulated, and that will allow you create different image weightings and optimize for time savings. While we note that there are also many forms of pulse sequences, and we are showing the basic form at this time[11], we also attempt to provide you a way to understand a pulse sequence through deduction and reasoning. But first, we start the next section with a fun and simple story.

4.3 Pulse sequence diagrams

There is an old folktale that we will invoke here:

Several wise blind men came across an elephant.

The first one felt the broad body, and said this is a 'retaining wall'... The second felt the tusk and said this is a 'spear'...

The third felt the tail, and said this is an 'asp'[12]... The fourth felt the ear and said this is a big 'fan'... The fifth felt the trunk and said it was a 'pillar' and so on....

---

[11] Beyond the basic pulse sequence structures that are presented here. There are many possible enhancements and advanced structures including EPI, FSE, diffusion prepulses and many others ones that are covered in references such as Brown *et al* [5] and Bernstein *et al* [6].

[12] Asp is a species of snake.

Figure 4.6. Invoking an old folk tale when we are discussing pulse sequence diagrams.

In this story, each man was sightless, unable to see the whole picture. Each describes a different aspect of the elephant, as depicted in figure 4.6. Even though each of these aspects can be considered separately, it is imperative to consider each man's understanding to establish the concept of the whole elephant.

In the same way, we must dissect the different parts of the MRI pulse sequence[13] in order to see the full 'image.' An MRI pulse sequence is a schematic diagram to describe the various parameters that, together, determine an image type. By altering just one of the parameters, the image will change. Therefore, we must consider each parameter both individually and as a part of the whole. It is only after they consider all of the parts together that they obtain an idea of the entire concept, in this case an 'elephant.' I hope this analogy provides a little fun, and at the bottom of this section

---

[13] This approach of dissecting a concept seems to be favored by radiologists, radiological technologists, and even learners from different areas, especially if they have limited study time for a single subject. But, this is not the only way, if you are a physicist and/or engineer who has a mathematics and physics background. I highly recommend *Magnetic Resonance Imaging* by Brown *et al* [5] and *Handbook of MRI Pulse Sequences* by Bernstein *et al* [6]. There are also a multitude of reasonable instructional videos that construct ideas from first principles on YouTube. These videos continue to grow in popularity and can be worth browsing.

Figure 4.7. Elephants are beautiful, and sometimes can feel mysterious. Although elephants can be both large and imposing, they can also be gentle creatures, and by looking carefully at their characteristics (like we propose to do with the PSD), the reader will begin to appreciate the uniqueness and intricacies of the underlying makeup of these creatures. Reproduced with permission from Sutipond Stock/stock.adobe.com.

we have a mnemonic for you to retain the lines of the pulse sequence using the elephant story metaphor.

Stop[14]! Before we can proceed to our safari[15] we should learn and observe some details that concern the structure of pulse sequence diagrams. You will be doing this through exploring parts of that 'elephant' in a deductive way as you proceed along this journey, as shown in figure 4.7.

We will next try to create the checklist of how to recognize the parts of a pulse sequence: Let's take a look at the pulse sequence, which is our 'elephant' that we want to explore. Have you heard of these terms?

So let's Try a Deductive Learning Sketch (DLS). What are the three Gradient Lines in this figure? Can you identify which is the **Slice**, **Frequency**, and **Phase** encoding gradient in figure 4.8?

---

[14] This Bilingualstopsign image has been obtained by the author from the Wikimedia website, where it is stated to have been released into the public domain. It is included within this article on that basis.

[15] We use the 'safari' (in the same way that the company 'Apple' uses the name in their web browser) to indicate the context of going out and observing nature. Looking around and learning things is a great way to get concepts from observation and deduction.

MRI: Connecting the Dots

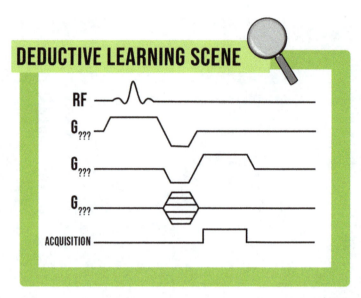

Figure 4.8. Note that G is a gradient and the reader is supposed to identify which line is the $G_{frequency}$, G_{Phase}, and G_{Slice}[16]. The reader is supposed to fill in the ?, ??, ??? above with the terms frequency, phase, and slice.

Note, this DLS is here to help expose the reader to the objective of the section. Don't worry! Like the elephant analogy, it's the goal that by the end of this section you will feel your way to understanding this topic. Some tips for learning these materials in class include divvying up the different lines of the gradients to different parts of the class so we can incorporate some peer learning/teaching into understanding the Pulse Sequence Diagram concepts[17].

First look at the deductive learning sketch and see if you can fill in the ?, ??, ??? example. The next few sections will go through the pulse sequence diagram above. If you were able to figure that out without looking at the answers, then congratulations, you may be already a head of a lot of learners on pulse sequence diagrams! However, if you couldn't figure them out, then the following few sections will hopefully clarify the PSD lines for you and what they do. We will use a peanut butter and jelly/jam sandwich analogy and the answers are: ?= slice, ?? = read, ??? = phase for the lines in figure 4.8.

NOTE: We've provided two metaphors here to help the reader relate to the potentially 'abstract' concepts. The entire pulse sequence diagram is the **elephant in that metaphor**

---

[16] One of the goals of this book is to move you on a journey to eventually better understand tradeoffs and artifacts in MRI, which we approach near the end of this book. At this time, a simple gradient echo pulse sequence diagram is used as the model for understanding the PSD. As mentioned earlier, there are many forms, and we hope the reader will seek to eventually learn and understand some of those forms in their future learning as they get a gist of the concept of slice, frequency, and phase encode for the present time of their learning trajectory.

[17] If you are an instructor, you can provide each student with the information prior to the class time, and then follow up with the interactive instruction during class.

that you need to conquer for better understanding. If the reader achieves the ability to recognize the 'lines' of the pulse sequence and their purpose, they have been successful in tackling the elephant. Additionally, a 'fun' mnemonic can be created to remember the lines by retaining the elephant metaphor. The next metaphor breaks down the gradient function. Each gradient (slice, frequency, and phase) is represented by a **sandwich metaphor** that includes the bread (slice), fresh jam (frequency), and peanut butter (phase). Perhaps a singular metaphor would have been simpler, but we give the reader some choice in how they want to structure their understanding (elephants and/or sandwiches)[18].

Let us set out to learn a few **goals** (collage shown in figure 4.9).

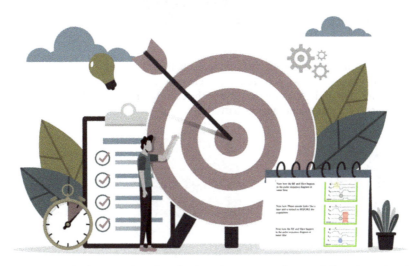

Figure 4.9. Goals for learning three lines in the pulse sequence diagram follow. Reproduced with permission from Diki/stock.adobe.com.

1. Connect that the slice gradient is timed up with the RF excitation pulse. In brief, it is the only line of the pulse sequence that does that. The reason is that it's the one that is tipping the spins required for precession and selects the slice with that gradient. (**Maybe this is the tusk/spear of the elephant/slice of bread**[19].)

2. Connect that the frequency gradient is timed up to occur simultaneously with the acquisition windows. The key point is this is like listening to radio station dials.

---

[18] If you are an instructor using this book, you may like to refer to one metaphor or the other, instead of both. We found that using both metaphors did not detract from learning, and hope this is the same case for your class.

[19] This bread slice image is reproduced with permission from rimglow/stock.adobe.com.

Now that you have a bandwidth of frequencies, you need a descrambler (the Fourier transform) to create your image in that direction. (**Maybe this is the ear [fan] of the elephant/fresh jam** [20].)

3. Connect that the phase gradient happens before the acquisition window. It looks like a ladder (note, we will see later in chapter 6 where each rung of this ladder is another step and increases the time of the entire sequence). (**Maybe this is the legs [pillars] of the elephant/peanut butter**[21]**.**)

4.3.1 Let's look at the 'tusk/spear' of the elephant, i.e., the slice encode

This where we will begin 'slicing' We can focus on the 'slice' line of the gradient. We will compare the slice gradient and its implications in relation to the RF function. Notice the gradient slice occurs at the same time as the RF excitation[22]. The slice gradient provides spatial localization (the location that you are selecting). Please try to hold in your mind that when you come across terms of **gradient**, think first if that gradient has a role in **spatial encoding**. As you will learn later, this is called the slice select gradient, as it localizes the region of interest of the slice that is to be excited. The goals of this section are shown in figure 4.10.

4.3.1.1 Language box: understand what a slice is
For some, because language is so vast, it is helpful to look at several word definitions to solidify these concepts to understand the 'slice' (shown in figure 4.11).
- What is a **slice**? MR image acquisition process subdivides a section of the patient's body into a set of slices;

---
[20] This jam image is reproduced with permission from baibaz/stock.adobe.com.
[21] This peanut butter toast image is reproduced with permission from Dan Kosmayer/stock.adobe.com.
[22] Recall that RF stands for radiofrequency, and in this case, an excitation refers to the transmission. We discussed radiofrequency transmission in section 2.4 in chapter 2.

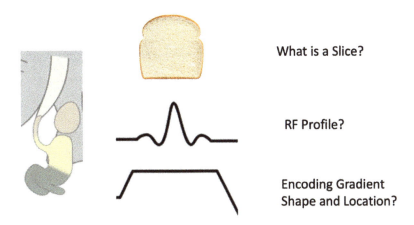

Figure 4.10. Collage with three targets for understanding the slice encoding gradient.

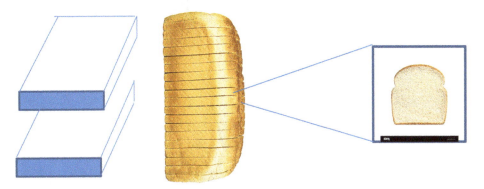

Figure 4.11. Slice selection and slices from a loaf of bread analogy. Note that this bread slice will have a slice thickness. Note the slice thickness is a specified parameter in an MR Sequence. For example, 4 mm may be a slice thickness parameter used in the brain. This loaf of bread image is reproduced with permission from danheighton/stock.adobe.com.

- A **slice** in MRI is a tomographic section of an object;
- A **slice** is defined by position and thickness;
- Each **slice** is a selection of spins in a plane through the object that will further divide into rows and columns (designated by a frequency and phase encoding).

Perhaps supplying four definitions is a little redundant, but we hope to break down the concepts in this chapter. The first goal is to cut down the section into pieces. For the simplest analogy: like the slices in a loaf of bread.

Here are some clues to proceed to process this DLS. Remember that the process of '**slice selection**' requires an RF pulse to activate and tip the spins in this slice, as seen in figure 4.12[23].

---

[23] Please note the shape of the RF pulse is sinc-shaped, a central hump with ripples on the side as shown in figure 4.13. The RF line in the pulse sequence diagram represents the RF Transmitter energy that is used to tip the spins, as we described in section 2.5.

Figure 4.12. RF pulse tips the proton spins, as we described in chapter 2 on the RF transmitter.

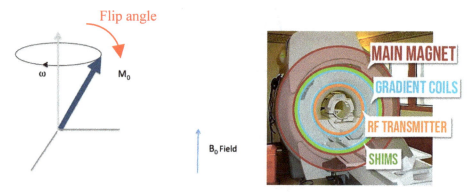

Figure 4.13. Tip angle is where the RF transmitter operates to create the flip angle. The flip angle is the tipping. Note that a gradient would be required to allow this to be spatially localized. The gradient pulse and the RF will be needed to be able to be operated together.

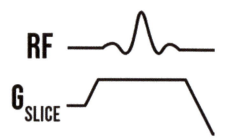

Figure 4.14. You can consider that the slice gradient provides slice spatial localization.

As a brief review, the energy tips the spins, as was discussed in chapter 2. Remember, it's the energy (magnetic field) that we imparted from the spins that causes the tipping. The RF pulse illustrates the amount of energy that the transmitter supplies to provide tipping of the flip angle, as seen in figure 4.13.

Remember, we need spatial localization!!! When we think of **spatial localization**, we should think about something to do with a '**gradient**.' If that is true, then slice selection will require the dual application of an RF excitation, as well as a gradient. Notice the gradient slice occurs at the same time as the RF excitation. The gradient slice provides spatial localization (the location that you are selecting), as shown in figure 4.14.

Finally, it is useful to see the slice selection in the context of the entire pulse sequence diagram, as portrayed in figure 4.15.

Now that you have recognized that the Slice gradient and the RF transmitter co-occur in time, the next step is to see how the tandem appears within the pulse sequence diagram (PSD) so that you can recognize how to find it in a sequence diagram.

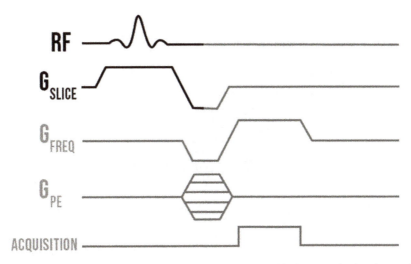

Figure 4.15. You can see the Slice Gradient Encoding Gradient along with the RF excitation shown within the context of the entire pulse sequence diagram.

Figure 4.16. You started with constructing a bread slice. You should celebrate understanding slice with jam.

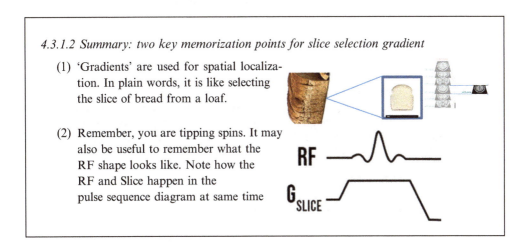

4.3.1.2 Summary: two key memorization points for slice selection gradient

(1) 'Gradients' are used for spatial localization. In plain words, it is like selecting the slice of bread from a loaf.

(2) Remember, you are tipping spins. It may also be useful to remember what the RF shape looks like. Note how the RF and Slice happen in the pulse sequence diagram at same time

Were you able to figure this out? Congrats! Give yourself some 'jam' to celebrate, as you can see in figure 4.16.

Figure 4.17. Continuing on your safari, you can feel your way around more of these elephants to gain more intuition (i.e., the pulse sequence diagram). Reproduced with permission from EcoView/stock.adobe.com.

Note, if you are enthusiastic about what you learned about this topic and at some time in the future want to learn more details, please go read on appendix A.1 to read more. But, at this time, see figure 4.17.

Carlie P. (Premed and Dancer) reflects on the slice selection mechanism and said:
"I think the way to tell if it's a slice selection gradient application is if the gradient is timed up (occurring simultaneously) with the RF pulse that the system is transmitting."

Dr. Wu responded:
"I think you are getting it. Also, we will go into more details in the first part of chapter 5. where we also go into more specifics on how to calculate the slice thickness from the RF transmitter bandwidth and size of gradient."

4.3.2 Looking at the 'ear/fan' of the elephant, i.e., the frequency encode

An overall conceptual picture for the reader is seen in figure 4.18. Let's provide a little background on frequencies (we covered this in chapter 1), which is shown in figure 4.19.

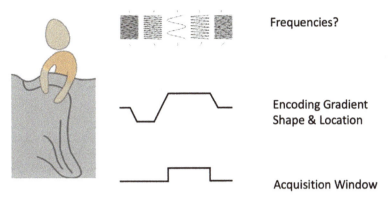

Figure 4.18. Three top concepts that you need to integrate to understand frequency encoding.

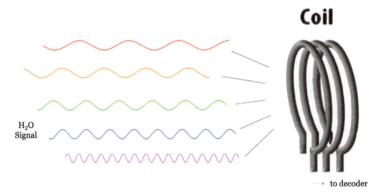

Figure 4.19. Figure representing the ensemble of waves that make up the signal. This is sent to the receive coil, which decodes the message as an image.

4.4 Scaffolding box: Steps of what is happening in frequency encoding.

Let's look at what happens in the scanner in a step-by-step process. In this case, we are focused on the just the frequency encoding direction for the moment.
- Water has protons. We know protons have resonance, so when tipped they can emit a signal (as a wave).
- Each of these signals are simultaneously acquired (which is related to the amount of water at a location) to create a 'packet' of signals.
- The packet the signals come from being summed together (i.e., intermixed), but travel to the receiver.
- The receiver sends the information to a decoder. We need math (the Fourier Transform to descramble the signal) to retrieve the encoded amplitudes, just like we did in the PHET interactive experiment (see section 1.1.6 of chapter 1).

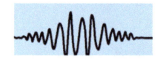

Figure 4.20. The encoding of the flour, water, yeast, and sugar make up bread. This is the analogy we are communicating that you are encoding information about parts of the object in the form of a wave to be sent. This encoded wave is encoded into frequencies, which we will later have to decode to produce an image. Reproduced with permission from New Africa/stock.adobe.com.

Now recognize that each location across the x axis will have a different frequency, for which we will encode the parts as an analogy shown in figure 4.20.

Perhaps this is a slight overgeneralization, but the image is made of many parts. This includes some things you don't even see, like the ingredients used to make that image. In the case of MRI, the ingredients are like the many 'waves.' Each wave has different frequencies[24] and amplitudes. We will want to extract the amplitudes from frequencies that we encode in. Those amplitudes can help us put together at least one of the directions in MRI, the frequency encoding direction.

At this point, we have a handle on the frequency encoding pattern. If you seek to delve into a more detailed explanation of the frequency encoding gradient, consider looking at the appendix 4.A.3. Next, we will help focus your attention on the appearance of the frequency encoding gradient in a series of parts of the pulse sequence diagram.

As we mentioned in section 4.3.2, you need to connect that Freq gradient is timed up to occur simultaneously with the acquisition windows, as seen in figure 4.21. The key point is that this is like listening to radio station dials. Now that you have a bandwidth of frequencies, you need a descrambler (the Fourier transform) to create your image in that direction.

As a walkthrough of our method of encoding up to this point:
1. Focus on the acquisition line as it relates to the frequency line. The acquisition line is not a gradient, but is the window through which we listen to the signal (this is a learning point), as shown in figure 4.21.
2. Recall that we discussed the frequency encoding in the applied bubble at the beginning of this section. We can see that time acquisition is happening at the same time as the frequency encoding. Spatial encoding while the frequency gradient is turned on is going to produce a spatial encoding related to frequency. I like to think of this as a radio station dial where you acquire different channels during the acquisition time.

---

[24] In section 4.3.5. in this chapter, we will also need to incorporate 'phase' in addition to frequency.

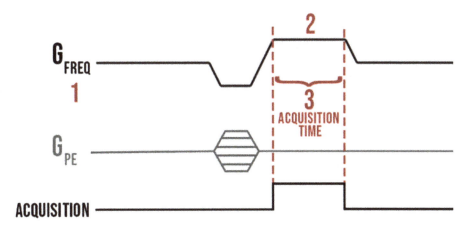

Figure 4.21. Illustration demonstrating the aspect of listening to the signal evolution in the presence of the frequency gradient being turned on.

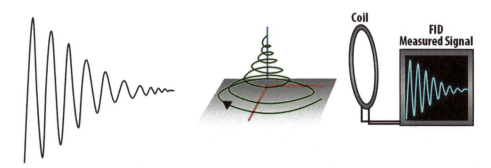

Figure 4.22. This figure shows in context the main frequency encoding lobe, as well as where it is situated with the data acquisition window.

3. For simplification, focus on recognizing that you are listening to a bunch of frequencies and their amplitudes for a period of time in the presence of a gradient. This gradient spatially encodes water in locations based on the amplitude (signal from protons). See chapter 1 section 1.1.9 and section 2.4. The Fourier transform unscrambles the 'message' and turns those values into an image.

See figure 4.22 for the frequency encoding gradient as well as the acquisition window in the context of the entire pulse sequence diagram.

The next step in creating an image is frequency encoding. By this point, we have the frequency pattern and the gradient. Now, we need to encode it. More details on the frequency encoding gradient are included in appendix 4.A.2 at the end of this chapter.

4.4.1 Summary: two key memorization points for frequency/read encoding gradient

(1) Note now that the **read encode** happens at same time as the **data acquisition window**.

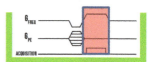

(2) Recognize how all the frequencies get pulled in and need to get descrambled to generate a profile.

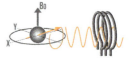

Nick P. (third-year medical student) reflected and provided some takeaways from reading about the frequency encoding direction: "Frequency encoding can be one of the most difficult sections to learn and understand. It combines several potentially abstract concepts. The first idea is that you are listening to a bandwidth of frequencies (a range of frequencies) that consists of different sets of waves with different frequencies. As we described above, you can encode the amplitude of a signal at a location into those frequencies. They can be transmitted over a far distance and then received by an unscrambler. Finally, they can be turned into a representation of an image. The data acquisition window determines which frequencies should be timed up with the actual frequency encoding gradient. The gradient itself also determines the spatial localization of that signal."

Were you able to follow and get the two points on frequency encoding? That is quite an achievement! Please continue spreading the jam[25], or add 'frequency encoding' to your list of accomplishments, and please continue your journey on understanding the PSD (with the elephant analogy as shown in figure 4.23). Note, if you are keen on the topic of frequencies, and perhaps some time in the future want to learn more, please go to appendix 4.A.2 to read more.

---

[25] The above jam bread image is reproduced with permission from Dan Kosmayer/stock.adobe.com.

Figure 4.23. Let's look at one more part of the elephant (another part of the pulse sequence). Reproduced with permission from sdbower/stock.adobe.com.

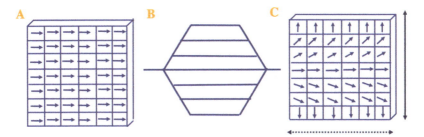

Figure 4.24. A) Block of encoding where rows illustrate signal prior to the gradient application, which has a phase of zero as shown by the arrow pointing to the left. B) Shows a simplified (reduced number of lines) phase encoding gradient block. Note that for each iteration, a different phase encoding level will be applied. C) Due to the level of the phase encoding encoding, then each row of the block will experience a different phase of encoding. Note that the solid double black arrow points in the direction of phase encoding, while the dotted black double arrow is pointing in the direction of frequency encoding.

4.4.2 Looking at the 'legs/pillars' of the elephant, i.e., the phase encode

Note that the phase encoding gradient and its shape and structure of that gradient looks like a ladder (figure 4.24). The next figure will illustrate the phase encode within the context of the entire pulse sequence diagram.

To visualize phase encoding it is instructive to refer back to how we described the 'wave.' Recall from section 1.1.6 that we described two parameters that we used to represent the form of the wave as 1) frequency and 2) phase. We have used frequency to encode one direction of the image. We can also encode a second direction (the phase-encoding direction). If you only vaguely remember what phase and frequency look like, it might be useful to go back to figure 1.12 and review the differences in the way waves appear. Figure 4.24 contains a visual representation of the phase encoding acquisition.

We discussed how to create image phase encoding in one direction. In this section, we will mention how we are able to encode in another direction. If you are dedicated to learning more about this topic at some time in the future and want to learn more details, please read appendix 4.A.3.

4.4.3 Summary: two key memorization points for phase encoding gradient

(1) Note now the **phase encode** happens just before the **data acquisition window**.

(2) Recognize differences in how different phases form.

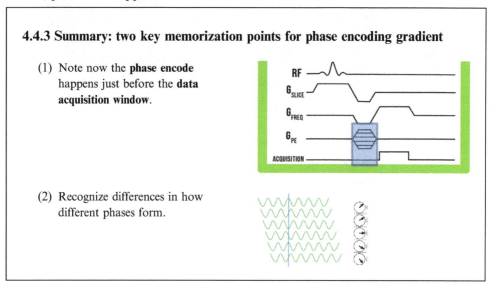

Now that you've seen how the phase encode works, please feel free to give yourself some peanut butter to go with that jam[26].

Dr. North reflected on phase encoding:
"I think phase encoding is important to know for artifact directions. Occasionally, I might also think about the time that a scan takes, but that seems to be managed more by our technologists."

---

[26] The two bread images are reproduced with permission from Dan Kosmayer/stock.adobe.com.

4.4.4 INSTRUCTOR TIP

For instructors, one fun way to possibly teach this concept is to form two or three groups of students. This will be a way to help learners just focus on one of the three concepts and help their colleagues understand the others. If you are only doing two groups (see figure 4.25), we'd suggest slice encoding for one group and frequency encoding for the second group. The instructor can just mention how phase encoding works and, more importantly, how to identify phase encoding in the pulse sequence diagram.

Provide each group each with the information prior to the meeting and then also add additional class time. For example, allow 20 minutes of the class time for students to discuss and then complete the teaching with summarizing the interconnected components.

Here's a possible way to create 2–3 groups (depending on how many students you have)[27]

- Have the first group work on **slice selection** and explaining the pulse sequence parts.
- Provide the second group with the information to describe how frequency encode works.

Figure 4.25. A fun way to do this is to break your learner group into smaller groups. Hard-working students are illustrated in this picture. Reproduced with permission from rh2010/stock.adobe.com.

---

[27] If you want, you can also send a video of how this works to your students prior to class. They can select which group they want to be in during the prior class or be randomly assigned. One example would be a video on frequency and radio station dials.

- If you have a third group, provide them with the pages on how the phase encode works.

Then, use the 30 minutes left in the class to have them explain it to each other the best they can, and perhaps 10 or 15 minutes per group for them to present and get feedback.
Finally, use 10 minutes (assuming you have an hour) to pull all the items together in a summary table. Also, if you are doing this in different years, repeat this exercise in years one and two.

4.4.5 Summary of parts of the PSD all together

Reproduced with permission from tonymapping/stock.adobe.com.

Here is a key summary (of summaries) and a little more detail on TE and TR in the pulse sequence diagram.

Final summaries for all three parts you have learned

Note how the RF and Slice happen in the pulse sequence diagram at same time.

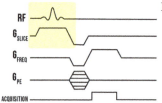

Like the Slice of Bread in our motivational analogy

Note how Phase encode looks like a ladder and is turned on BEFORE the acquisition.

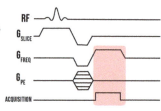

Like the Jam in our motivational analogy

Note how the RF and Slice happen in the pulse sequence diagram at same time.

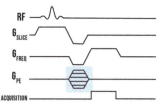

Like the Peanut Butter in our motivational analogy

4-25

As you have made your way through the three parts of the pulse sequence diagram, you have certainly put a lot of pieces together, and that takes a lot of 'heart' (see figure 4.26). Please, continue your journey forward, and don't forget to celebrate your journey into MRI. You've made good progress on your safari.

Figure 4.26. An image of fresh jam added with peanut butter on bread as a visual reward for understanding three parts of the pulse sequence (frequency and phase encode and slice encoding). Reproduced with permission from Dan Kosmayer/stock.adobe.com.

Figure 4.27. An elephant analogy for understanding the pulse sequence diagram. Learn the diagram piece by piece as you feel your way through each line.

If you've accomplished the above task and can restate the labels of all three major gradient directions and are enthusiastic about what you learned, then sometime in the future, please go read on appendix A.4 to read more as practice in consolidating your knowledge.

In summary, you have now 'felt' your way through the pulse sequence, which we hope is something exciting and fun, and considered the analogy of the elephant, as shown in figure 4.27. Pulse sequences are at the heart of MRI software and 'playing' out these pulse sequences and their gradients provide the unique contrasts that we rely on every day for clinical diagnoses.

Now that we have walked you through a 'pulse sequence' diagram, here is the mnemonic for recalling the elements of a pulse sequence by remembering the story of the elephant:

The first felt the body, and said this is a 'retaining wall' (R = RF Transmit)
The second felt the tusk, and said this is a 'spear' (S = Slice)
The third felt the tail, and said this is a 'asp' (A = Acquisition)
The fourth felt the ear, and said this is a big 'fan' (F = Frequency)
The fifth felt the leg and said it was a 'pillar' (P = Phase Encode)

Finally, if you have made your way through the pulse sequence diagram you now have access to knowledge on the construction of how the MRI scanner runs. Well done! The concepts of relaxation and the pulse sequence diagram are central to many of the final chapters in this book that deal with tradeoffs in signal and time (as well as heat deposition), and artifacts, which are not only testable concepts, but also aid practitioners with MRI throughout their careers.

4.4.6 Scaffolding bubble on TE and TR

Two of the most important timing terms to learn in the pulse sequence are shown in figure 4.28. The first is repetition time (TR), which is the time from beginning to end, starting approximately with excitation from the RF pulse. The pulse sequence block is what is repeated for each RF pulse. (In some cases, it can be thought of as the time between each initial excitation pulse for the purposes of T1/T2 contrast.)

Let's move on to the last part of this chapter, which concerns the 'amazing' and useful spin echo [7][28].

4.5 Spin echo

In this section, we discuss how a spin echo pulse sequence partly corrects for inherent artifacts. Distortion and signal loss arise due to local inhomogeneity between different materials and their response to magnetic fields (such as magnetization

---

[28] Spin-echo methods are less affected by signal loss and distortions from field inhomogeneities and tissue-induced susceptibility variations [7].

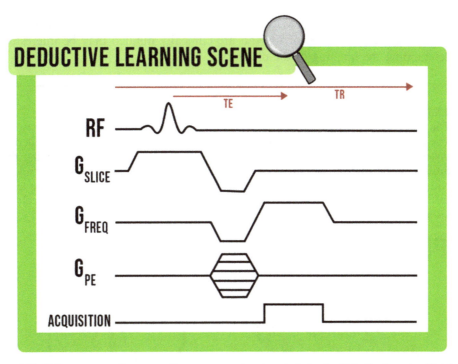

Figure 4.28. Pulse sequences. For many of you, you are almost ready to try the Detective Learning Sketch. If you are already able to make progress explaining this figure to others, kudos to you. We are impressed that you have been that intuitive and have put together concepts that will help you be able to evaluate other advanced sequences (like EPI and FSE). Otherwise, please read on, as we will slowly try to help you appreciate the components of this elephant (i.e., the pulse sequence).

effects that occur between air, bone, and tissue interfaces)[29]. This inhomogeneity is due to areas of varying magnetic susceptibility, which corresponds to the internal magnetization of a tissue when placed inside an external magnetic field. To understand this further, please consider the following learning objectives:

Learning objectives:
(1) Spin echo is typically implemented to reduce effects of local inhomogeneity.
(2) Composed of two subsequent RF pulses, the first at 90° and the second at 180°.
(3) Spin echo is considered slower than gradient echo in terms of time taken to acquire.
(4) Can provide T1, T2, or PD contrast.

Questions to consider:
- Explain the concept of T2* and describe its general relationship to T2.
- Describe and illustrate how a 180° RF pulse produces rephasing of protons.

---

[29] Recall the discussion surrounding T2* in chapter 3 that also describes the bending of waves at material interfaces.

A potential place to start is to look at an example of where we might suggest using spin echo sequences. Those of us who are clinically minded may be interested in understanding the imaging of the pituitary.

4.5.1 Applied bubble: pituitary gland

The pituitary is a small hormone-producing gland that is located behind the nose. The pituitary, paired with the hypothalamus, controls all of the glands in the endocrine system, including the thyroid and adrenal gland. MRI is useful for examining the functional status of the pituitary gland and evaluating pituitary lesions, such as micro and macro adenomas. MRI is used to evaluate the pituitary as the organ is made of soft tissue, and also because the bony structures around it make it more difficult to visualize with other modalities, as depicted in figure 4.29.

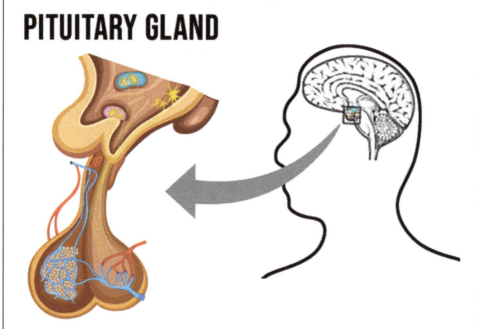

Figure 4.29. A visual picture of the pituitary, which plays a central role in the neuroendocrine interface. Adobe Stock Images: © vecton.

Recall susceptibility and the T2* effect. These effects consider the EM receptive nature between materials. The consequence of the magnet field changes as it passes through different regions of the body (as we described in the T2* portion of chapter 3) creating a bend in field, which further results in signal loss and/or distortion, which we describe as effects due to local inhomogeneities. These concepts are important in understanding the spin echo, so first we are going to examine a DLS, which is shown in figure 4.30.

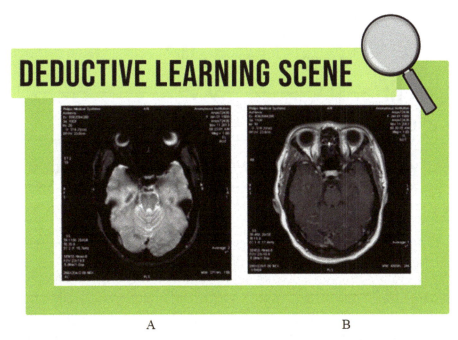

Figure 4.30. (a) Gradient echo and (b) spin echo provide different 'looks' when focusing on the pituitary gland.

Compare the differences between image A and B. Image A is a gradient echo, while image B is a spin echo. In which image is the pituitary gland more clear? What we want to point out is the susceptibility effect on signal loss, particularly in the pituitary region (behind the nose, proximal to the optic chiasm [the optic chiasm position is slightly superior in the head than the slices shown]).

The pituitary is better visualized on the (B) spin echo acquisition (i.e., less distorted and less signal loss than in (A) gradient echo). The reason that this distortion arises is because you have magnetic fields passing through an air-to-tissue interface that causes the waves to bend and results in signal loss and spatial distortion.

Let's look at the spin echo pulse diagram as shown in figure 4.31. We previously described the effect of local inhomogeneity, so you may now observe the reduced effects of local homogeneity due to the use of the spin echo pulse sequence.

Notice the multiple pulses. The spin echo pulse sequence has two RF pulses. The 90°–180° combination produces a 'refocused' signal known as a spin echo. The 180° RF pulse appears to have about twice the area under it (i.e., about double the size) and is known as the refocusing pulse. The 180° rephasing pulse compensates for the constant field heterogeneities to obtain an echo that appears to be T2-weighted and no longer T2* weighted (if you sample at the appropriate echo time). It is instructive to look at the pattern of motion of the spin echo that is able to 'refocus' the inhomogeneities, as shown in figure 4.32.

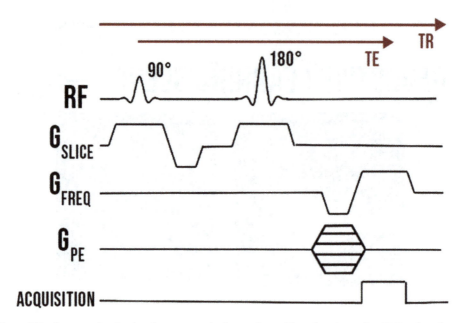

Figure 4.31. A conventional spin echo sequence is shown above. Note there is a pair of RF pulses shown in this pulse sequence, as opposed to the single RF pulse of figure 4.28. Note that we discussed spin echo in section 3.3 of chapter 3.

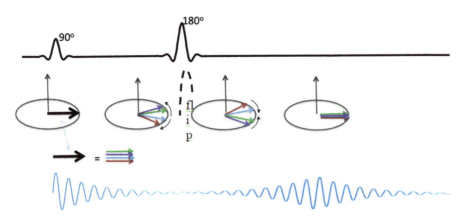

Figure 4.32. At time 0, the longitudinal magnetization is tipped into the transverse plane and begins to precess. In the presence of inhomogeneity, we can model these spins as having slightly different rotational rates, some faster and slower than the precessional frequency. The 180° pulse will flip these spins, moving the faster moving spins to the back, while the slower spins to the front. The spins now continue forward to rotate at a period that is timed to generate the conference of these spins, which will create the spin echo.

4.5.2 History bubble: runners on a track analogy for the spin echo

A favorite way to describe the spin echo effect is through the analogy of racetrack runners. We will let the description as written by Hahn from 1953 demonstrate the concept on its own [8] (figure 4.33) 'The echo effect can be explained from a very simple

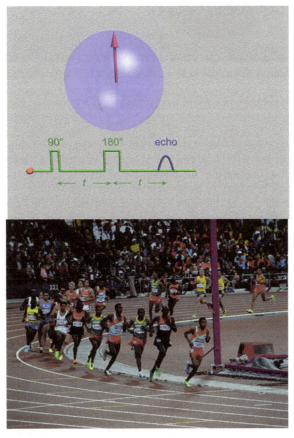

Figure 4.33. Hahn's famous spin echo analogy, which describes the signal as runners around a track that reverse direction and form a 'echo' in the signal [8]. This 'File:HahnEcho GWM.gif' image has been obtained by the author from the Wikimedia website where it was made available by User:GavinMorley under a CC BY-SA 3.0 licence. It is included within this article on that basis. It is attributed to User: GavinMorley. Animation available at https://doi.org/10.1088/978-0-7503-1284-4.

analogy. Let a team of runners with different but constant running speeds start off at a time $t = 0$ as they would do at a track meet (see the cover of this issue). At some time TI, these runners will be distributed around the race track in apparently random positions. The referee fires his gun at a time $t = T > T$, and by previous arrangement the racers quickly turn about face and run in the opposite direction with their original speeds. Obviously, at a time $t = 2T$, the runners will return together precisely at the starting line.'

Figure 4.34. Animation of spin echo, which can be helpful for visualizing multiple spins that have slightly different 'frequency' rates due to field homogeneity. This is refocused through use of the spin echo [10].

There are two major signal formations that are often 'sampled' during pulse sequences: the spin echo and the FID[30]. If you are clinically minded, i.e., technologists and/or radiologists, then a fundamental understanding of the free induction decay (FID) and spin echo (SE) will be useful to progress your mastery of MRI.

An animated version of the spin echo is shown below. This can be helpful for visualizing the motion, as shown in the video link provided in figure 4.34.

Finally, it is important to note throughout that as we described in the previous chapter, the spin echo only corrects for the field inhomogeneities term $1/T2^* = 1/T2 + \gamma \Delta B$. However, intrinsic spin–spin relaxation decay is still in effect, even with use of refocusing pulse, so the intrinsic molecular motion component (i.e., spin–spin relaxation component) will not be corrected by the spin echo, and only the $\gamma \Delta B$ term gets corrected.

Gail H. (neuroscience postdoctoral researcher) asked:
"How is spin echo different from gradient echo?"

---

[30] The spin echo and the FID are primarily forms of MRI signal that are discussed above. However, there is also is a third echo variation called the stimulated echo, which is beyond the scope of this book, but we provide a reference for stimulated echo as described by Frahm *et al* [9].

Dr. Wu replied:
"Essentially, the spin echo can refocus local gradient changes. Imagine you have two substances that are each of a different material. As you pass a wave through the substances, the wave will bend, much like light in the index of refraction model that you might have learned about in school. That bending is equivalent to magnetic fields in a local and unmeasurable value gradient. The spin echo reverses the field direction and enables us to recover part of that signal. While not all the signal is recoverable, the spin echo enables better recovery than the gradient echo. If you would like to read more about this and are interested in the conceptual details, please proceed to the appendix of this chapter."

Finally, we note that the spin echo can provide T1, T2, or PD contrast. There are some advanced pulse sequences that are optimized for speed, contrast, or signal to noise. A good review of the topic is described by Plewes in [16].

4.6 Summary

In this chapter, we first discussed the free induction decay (FID) as an introduction to the ideas of pulse sequence. Then, we proceeded to discuss what many of you were waiting for: a way to understand basic pulse sequences (which can help you recognize other complex pulse sequence structures). If you were able to walk through the steps of pulse sequences, then I believe you have done well. If not, keep trying! If you will be reading MR images in your future, then mastery of this topic would definitely be enhanced.

There is also the echo time (TE). Note after the initial RF pulse, there will be FID (exponential decay, see chapter 3). TE is the time between the center of RF to the middle of the acquisition window.[31] It is where we expect the strongest signal and it is typically in the middle of the read-out window. Two general classes of sequences are a spin echo (covered later in this chapter) or a gradient echo (which we showed in figure 4.28), and provide a method of refocusing that creates an 'echo' to be placed in the acquisition window, as shown in figure 4.31.

Next, we discussed a very important concept called the spin echo. One of the key points of the spin echo is to recognize what the impact is, especially in regions where there is the potential for susceptibility. There are tradeoffs in spin echo use (it can be slower), but it can be an important tool in some cases where substances exhibit moderate to large susceptibility differences.

The story of the elephant and the blind men is an old story, but one full of wisdom. We hope that this DLS helped inspire you to understand pulse sequences. Regardless, you have learned a small mnemonic that may create a fond remembrance that is something that you can take along with your journey into MRI.

---

[31] The center of the acquisition is idealized, but sometimes in some advanced pulse sequences, the echo is placed earlier in the acquisition window to create a faster acquisition.

We provided a brief and conceptual understanding of many of these topics, but certainly one of the most important things is to remain curious and interested. There are many online videos that have hours of content, as well as books that construct from first principles many of the ideas of pulse sequences, and I'd encourage you to also seek these out. I've found many videos online that teach MRI with respect to specific body parts and pulse sequence choices have direct implications to the implementation of these clinical protocols. In these videos, I'm sure more of a foundation in the conceptual part of the acquisition will further aid you in this journey.

Appendix 4.A

Initially, chapter 4 was drafted to include greater details for the slice, frequency, and phase encoding lines that are part of the pulse sequence. However, after feedback on these concepts from a variety of residents, medical students, psychology students, and technologists, we decided to move some of the more involved physics/design of pulse sequences to this appendix. The purpose was to assist with the chapter flow and to focus the readers. However, it was anticipated that some of the readers would be additionally curious, particularly about some of the nuances in the profiles and appearances of the gradient lobes within the PSD.

For this appendix, we will break it into
1. Slice select additional nuances
2. Frequency encoding nuances
3. Phase Encoding nuances
4. Additional summarizing pulse sequence review

4.A.1 Slice select nuances (including refocusing gradient of the slice selection gradients)

Some readers may be curious about why there is an additional lobe after the main slice encoding gradients. You might be asking what is the purpose of the negative pulse in the slice encoding gradient. This is called the slice-rephasing lobe, as shown in figure 4.A.1. In brief, the rephasing corrects for the phase dispersion that occurs with the initial tipping of the magnetization concurrent application first slice-select gradient. Without the following rephasing lobe, phase dispersion leads to signal loss [12].

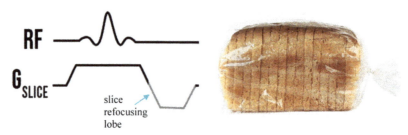

Figure 4.A.1. RF and slice encoding gradients highlighting the refocusing gradient that follows. Right image reproduced with permission from lufter21/stock.adobe.com.

4.A.2 Frequency encoding gradients nuances

Next, we need to look more at the frequency encoding gradient. We will first take a closer look into greater detail in the frequency encoding gradient line of the pulse sequence diagram. Please note that this gradient when depicted as a constant is a slope of changing magnetic field per change in space ($\Delta B/\Delta x$), as seen in figure 4.A.2. Also note that to achieve this gradient level (i.e., have turned up to a level) and/or turned off a level, a transitional slope must be used, as it is impossible to make discontinuous transitions in magnet field levels without time (see figure 4.8). Finally, as shown in figure 4.9, some readers may be curious about why there is a dephasing gradient pulse that is prior to the readout. This is because the spins that are to [32] be 'read' by the acquisition window are dephased in a way and then brought back together by the readout gradient.

In this appendix, let us take an in-depth look at the frequency line, dissecting each part and connecting our previous concepts, specifically gradients and spatial encoding. Pulse sequences are read from left to right, as progressions of time (figures 4.A.2–4.A.4).

Figure 4.A.2. A constant 'plateau' in this figure indicates a time where a constant gradient (slope) is provided by the MRI scanner. A ramp in these figures are the change from one gradient level to another gradient level which is known as gradient 'slew.'

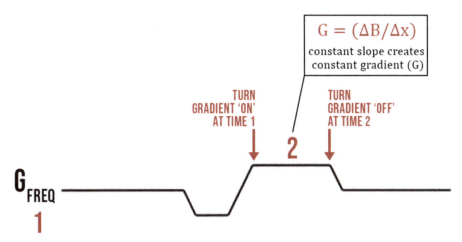

Figure 4.A.3. The constant gradient value corresponds to a constant slope of the magnetic field with respect to space. When discussing gradients in MRI, think 'spatially encoded'.

---

[32] Remember that gradients as we discussed in chapter 2 reflect a slope in the magnetic field per distance. Spatial encoding is performed through the application of the gradient. Note in the pulse sequence diagram, we are showing the amplitude of that slope over time. The changing gradients are timed in a way that assists with creating the 'weighting' or 'look' for that sequence.

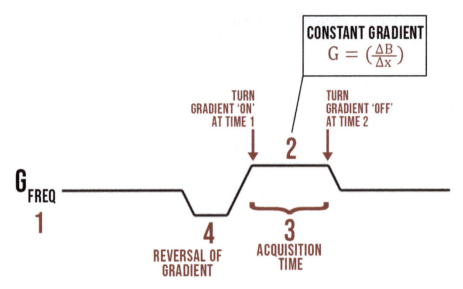

Figure 4.A.4. An annotated version of the frequency gradient (enumerated 1–4) as shown in the pulse sequence diagram. (1) note the 'freq' word identifying that is a frequency gradient if it is labeled, (2) is the constant gradient portion of the frequency lobe, (3) is when the signal is we are conducting the data acquisition, and (4) is the new part, which is the reversal gradient that may look unusual. The next section describes the purpose for advanced learners who may be interested in details on pulse sequence diagrams.

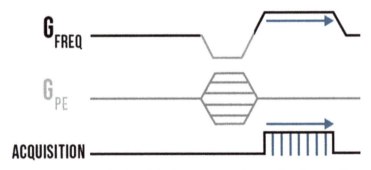

Figure 4.A.5. The figure shows the signal and the frequency encoding gradient in a gradient echo sequence (GRE) sequence. Note that the initial drop is the FID signal. This is coupled with a defocusing lobe, followed by a refocusing lobe that appears to be twice the area of the defocusing lobe. This centers the desired echo signal (known as a gradient echo).

(1) Notice the word 'FREQ' indicates that this gradient line is the (spatial) frequency, as shown in figure 4.5. Spatial frequency will be used to encode the signal using waves with different speeds of oscillation in space (i.e., different frequencies). A pictorial representation of this event is shown in figure 4.A.5.

(2) The word 'constant' refers to the slope with respect to space of the gradient $\left(\frac{\Delta B}{\Delta x}\right)$. The change in magnetic field will change with respect to the distance in space. Note that this is the constant value of the 'slope' that is turned on in time in the

above diagram, as this can be a point of initial confusion, and separating the ideas of space and time in this diagram should be carefully understood by the reader when looking at these pulse sequence figures.

Note that this frequency is turned on for a time, and during that time, the system starts to accumulate area under that curve.[5] This is a preview of what is the complicated topic of Fourier transform. We have introduced the Fourier transform[33] and k-space at later points in the book. Note that you are applying this constant gradient for a time in which you are also listening to the signal, as shown in figure 4.A.5.

These concepts can be pulled together. The more proficient readers will recall chapter 2's DLS exercise on $\Delta\omega = \gamma G \Delta x$. If you substitute $G = (\Delta B/\Delta x)$ into $\Delta\omega = \gamma G \Delta x$, then $\Delta\omega = \gamma \Delta B$. Continue to review chapter 2 if you would like to re-examine more details in the appendix of that chapter. Finally, you may notice the relationship similarities with the Larmor equation at this point.

Here we present one final concept that can help with additional understanding. We briefly introduce the concept of sampling frequencies in figure 4.A.5. At different times in the interval of the acquisition window, if you were to bin/chop up parts of that window, you would sample different frequencies. At those bins of time (i.e., sampling), you would be able to retrieve different frequency components. The area under the gradient contributes to the phase of the image. While you are accumulating phase, you are effectively evaluating it at various sampling intervals. Each of these corresponds to a different frequency measurement and may be useful to conceptualize, as seen in figure 4.A.5 [5].

Traveling across the frequency gradient is similar to scrolling through the radio. As you scroll the radio dial, you listen to each station briefly. Each station tunes in at a different frequency. As we move along the frequency gradient, signal is acquired at subintervals during a period known as the 'sampling time.' Each subinterval provides a different sample of the spatial frequencies. The sum of acquired frequencies is then turned into an image through the use of the Fourier transformation, which will unscramble the content that was encoded by the frequency encoding line.

Sampling is an advanced topic beyond the scope of this book, with the exception that it will be briefly introduced in chapter 5, when we explain receiver bandwidth. There are plenty of very good resources that delve into this matter, including mriquestions.com [13] and a video by Joseph W. Owen, MD [14]. The author believes that some of the best videos to learn these materials are from other medical professionals, as they are in tune with some of the direct learning experiences by radiology residents and technologists.

4.A.2.1 The negative lobe of read gradient [5, 6]

Advanced learners may be curious why there is a negative portion prior to the positive lobe of the gradient refocusing lobe, as illustrated in figure 4.A.6. The negative portion is a reversal in frequency, known as the negative gradient lobe [5, 6].

---

[33] The Fourier Transform (FT) for the purposes of this book can be thought of a tool to unscramble encoded waves that are used to produce an image. If more information about the FT is sought, MRIquestions.com [11] has a helpful webpage and 3BlueBrown has a nice video to augment your understanding [4].

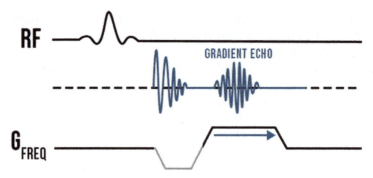

Figure 4.A.6. The acquisition window represents the duration of sampling. During this window, many frequencies are acquired at subsequent intervals (blue lines).

It is not crucial to every learner's understanding, but it is a question sometimes discussed occasionally between diagnostic radiologists and technologists of the MRI physicists. This section describes some intuition in regards to that negative lobe for those who are interested in better understanding parts of the pulse sequence diagram. A gradient reversal 'refocuses and centers' the spins, resulting in an echo-like effect. The reversal is used to improve image quality by creating greater symmetry in the acquisition, allowing better image reconstruction.

4.A.2.2 Advanced discussion of frequency encoding

Note the value of the frequency gradient determines the slope of the field with respect to space, while the acquisition window is the duration of time when the slope is turned on and off, as shown in figure 4.A.6[34]. The rephasing lobe allows spins to be rotated by the dephasing lobe and by contextually scrambling and then unscrambling spins, then pulling these back together to form the final more refocused signal. This is called gradient refocusing and will follow a T2* decay pattern. A nice description that we find useful for some advanced radiology residents is found at mriquestions.com [15]. Please continue to work on these understandings, as they can play a role in your comprehension of other and more complex pulse sequence diagrams in the future.

4.A.3 Phase encoding dissected a little more

There are a few concepts that we would like to reinforce for the phase encoding gradient. At this point, you should be able to recognize the shape that appears as a series of almost what looks like a ladder with rungs, as seen in figures 4.A.7 and 4.A.8.

Now we proceed to evaluating the phase encoding line. A gradient is applied before acquisition to slightly alter the phase of the acquired signal. The entire pulse

---

[34] Because of the multiple levels of understanding in phase, frequency, and Fourier transform, we do not have a straightforward way to describe this aspect concisely without some mathematical tools. Also, it demands someone to be very interested in the details. We thus refer you to excellent texts and websites, including [5], for more ideas.

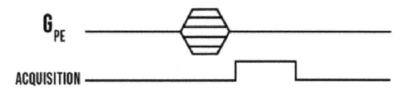

Figure 4.A.7. Note the representational appearance of the phase encoding gradient at the top of this figure that illustrates the pattern. Note that this is reflecting many lines of the phase encode, but represents many steps compressed into a simplified representation. One step may look like figure 4.A.8.

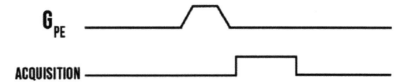

Figure 4.A.8. Note the appearance of a single phase encoding gradient line. This figure is provided as a convenience for readers that are not using the e-book version. The ladder allows the embedding of the animations directly.

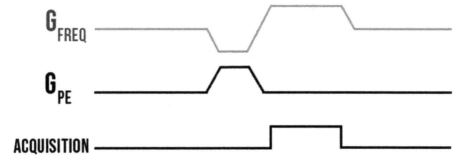

Figure 4.A.9. Phase encoding is similar to a ladder. Each 'rung' of the ladder indicates a phase change. The pulse sequence is repeated for every phase change. Animation available at https://iopscience.iop.org/book/mono/978-0-7503-1284-4.

sequence is repeated multiple times. Each time, the phase encoding gradient changes to provide spatial encoding. Note that the concept is revisited in chapter 6, when we describe k-space in more detail. The main 'bits' of phase-encoding may feel relatively easy to conceptualize, but it takes several tries to wrap one's head around the entire concept [11]. Please note the animation version, as shown in figure 4.A.9.

4.A.4 Overall review of pulse sequence appearance

Finally, in this chapter we provide an appendix that permits the reader to practice one more time with identifying the lines that are in a pulse sequence diagram. We start by enumerating them in the below figure and leave the chapter with one final TLS so that readers can practice and gain confidence in their ability to identify the lines of the pole sequence diagram. As a reminder, it is useful to understand the different encoding directions in MRI because they impact the time of the sequence and the signal-to-noise, as well as impact artifacts.

Let's break it down line by line. I've numbered the lobes (parts) in figure 4.A.10 to help you get started with the exercise. The negative rephasing lobes[35] are shaded gray, as we will focus on the main content of the pulse sequences.

Note that I highly recommend that you work on this in a team to begin to understand the pulse sequence.

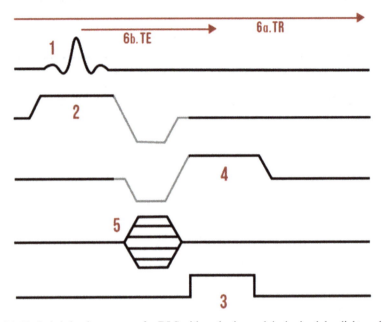

Figure 4.A.10. Labeled pulse sequence for DLS with rephasing and dephasing lobes lightened so that the reader can focus to figure out piece is associated with each number Acquisition window, G_{freq}, G_{PE}, G_{slice}, RFTransmit, TR, TE. Try to identify each line without looking back at the chapter, if you can.

To enumerate the answers to the DLS in figure 4.A.11.
1. RFTransmit
2. Gslice
3. Acquisition
4. Gfreq
5. GPE
6a. TE
6b. TR

If you have completed this exercise and truly can identify these sort of structures on a pulse sequence diagram—tremendous work! With this understanding, you might be at the top of your classes in clinical/translational MRI physics.

You have achieved a degree of mastery worthy of the elephant now being able to get your peanut butter and jam sandwich. Congratulations on your outstanding effort!

---

[35] A brief description of the refocusing lobe is provided in the scaffolding bubble above this section for those further interested. More details and links are included in the appendix.

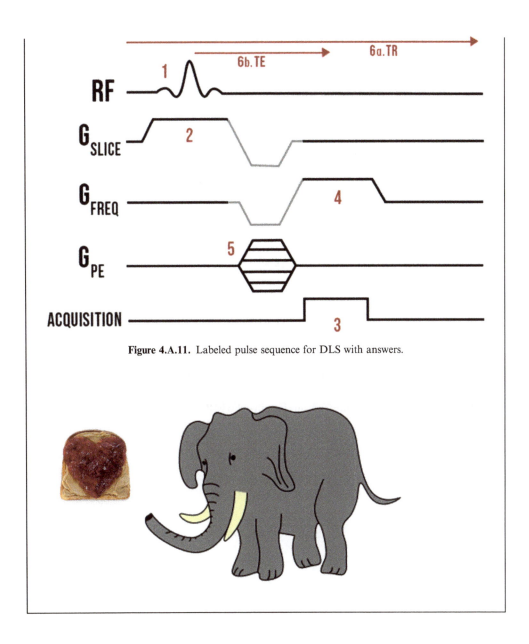

Figure 4.A.11. Labeled pulse sequence for DLS with answers.

References

[1] The History of Hello World 2015 *The Software Guild* (https://www.thesoftwareguild.com/blog/the-history-of-hello-world/)

[2] Kernighan Brian 1974 Programming in C: A Tutorial (PDF). Bell Labs. Retrieved 9 January 2019

[3] magritek. 2009, June 8. *Introductory NMR & MRI: Video 04: Acquiring a Free Induction Decay (FID)* (https://www.youtube.com/watch?v=MPXbDDRumwM)

[4] 3Blue1Brown. 2018, January 26. *But What Is the Fourier Transform? A Visual Introduction* (htttps://www.youtube.com/watch?v=spUNpyF58BY)

[5] Brown R W, Cheng Y-C N, Haacke E M, Thompson M R and Venkatesan R 2014 *Magnetic Resonance Imaging: Physical Principles and Sequence Design* 2nd edn (New York: Wiley)

[6] Bernstein M A, King M F and Zhou X J 2004 *Handbook of MRI Pulse Sequences* (Oxford: Elsevier) pp 267–97

[7] The AAPM/RSNA Physics Tutorial for Residents. Contrast Mechanisms in Spin-Echo MR Imaging. | RadioGraphics. n.d. (https://pubs.rsna.org/doi/10.1148/radiographics.14.6.7855348)

[8] Hahn E L 1953 Free nuclear induction *Phys. Today* **6** 4–9

[9] Frahm J, Merboldt K D, Hänicke W and Haase A 1985 Stimulated echo imaging *J. Magn. Reson.* **64** 81–93

[10] kas pijpers. 2010, April 4. MRI principles 4 of 4—180 degree pulse.avi. https://www.youtube.com/watch?v=GDElT6Tz7_Q

[11] Phase-encoding. n.d. *Questions and Answers in MRI* (http://mriquestions.com/what-is-phase-encoding.html)

[12] Elster A D 2021 Slice-select rephasing (http://mri-q.com/ss-gradient-lobes.html)

[13] Receiver bandwidth. n.d. *Questions and Answers in MRI* (http://mriquestions.com/receiver-bandwidth.html)

[14] Owen J W 2020 January 6 Radiology education *MR Physics 5—Bandwidth*. https://www.youtube.com/watch?v=P9NEr-ljp9c

[15] Gradient Echo (GRE). n.d. *Questions and Answers in MRI* (http://mriquestions.com/gradient-echo.html)

[16] Plewes D B 1994 Contrast mechanisms in spin-echo MR imaging *Radiographics* **14** 1389–404

IOP Publishing

MRI: Connecting the Dots
A start to concepts
Dee Wu

Chapter 5

Getting serious with MRI

(Further concepts into transmit, spectral components, and receive)

Congratulations! Chapters 3 and 4 provided you with a solid conceptual start to MRI, including relaxation, the pulse sequence diagram, and the spin echo, as shown in figure 4.1. If you have mostly mastered these concepts in those chapters, congratulations! You have achieved some solid fundamentals that we can build upon. Next, we need to focus on the three key items shown in figure 5.1. It is now a matter of mastering details bit by bit. This chapter will focus on the following items:

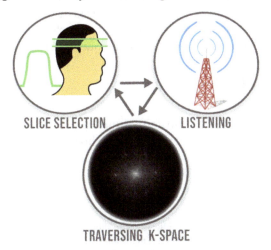

Figure 5.1. An overview of all of the above parts of this chapter. Note that the three items can happen in a sequence, but there are times when they can overlap in the timing of their application[1]. Tower image from Adobe Stock Images: © woverwolf.

---

[1] For example, you may be listening while the gradients traverse k-space. RF excitation can occur prior to listening to a signal, or can be part of the creation of a spin echo. In this way, they are all part of the MRI pulse sequence toolkit that provides you with the MR image.

A. Excitation and **slice selection.**
B. ***K*-space properties.**
C. Acquisition in regards to **signal-to-noise ratio** (SNR) and the **receiver bandwidth**.

The three topics mentioned above are part of concepts that we initially began to describe briefly in chapter 2. However, we felt it was important to go over these topics in more detail through this additional chapter, after the reader had greater exposure to and understanding of pulse sequence design (chapter 4). These topics will include: 1) **Slice selection**, which is related to RF transmitter of section 2.5 and that works in coordination with MRI gradients of section 2.4. The main 'job' of the MRI gradients is spatial localization. 2) However, a second and important role of MRI gradients is how these gradients help the pulse sequence transverse '*k*-space.' You were briefly exposed to this concept in section 2.7. It is now time to go into further detail on that topic, as it is useful when it comes to artifacts and advanced pulse sequence design, as the traversal will dictate the image weighting, timing of the sequence, and the reduction or enhancement of artifacts. 3) Finally, the **receiver bandwidth**[2] is a key concept to learn about, as the receiver bandwidth controls the signal-to-noise ratio of an image through the amount of noise that is allowed into the receiver by judicious choice and length of the acquisition window.

The ideas we discuss in this chapter are what you may encounter day-to-day in a busy radiology practice. The fundamental concepts not only cover the relaxation and pulse sequences that we previously discussed, but also build upon those and the manipulation of parameters that accentuate the contrast in terms of T1 and T2, in particular in the presence of variation and noise. (Remember there are more concepts, but we will primarily consider these ideas at the moment to provide you with a chance to master them.) As we proceed in this chapter, we hope that you are able to build fundamental skills and improve your understanding in these topics.

5.1 Excitation with RF transmit[3]

Learning objectives:
1) Understand that the RF transmit frequency specifies the energy range that is involved with the slice width excitation shown in figure 5.2.
2) Bandwidth of the RF transmit signal determines the range, or spatial localization, of the slice dimension. A smaller bandwidth produces a smaller slice.

---

[2] We want to stress when you see the word 'bandwidth,' you should recognize that bandwidth is only a general term that indicates a range of frequencies. It is very important that you qualify to which set of frequencies you are referring. Even though this may be slightly redundant because of confusion by learners, it is valuable to restate it here. There are two different uses that will define the context; one is for the transmitter and the other is for the receiver. The first is transmitter bandwidth, which is associated with the slice thickness, as we will see in section 5.1. The other is the receiver bandwidth, which is concerned with the acquisition window size and influences the Signal-to-Noise Ratio (SNR) in an acquisition.

[3] Remember how in chapters 2 and 4 we described how we created spatial localization in the frequency and phase directions. This section deals with how we create slice encoding, such as designating the size of slabs or slices of an image.

TRANSMITTER   **RECEIVER**

Figure 5.2. RF Transmit occurs over a bandwidth of frequencies. For MRI, we typically operate in a range of RF that is centered on the Larmor frequency with a bandwidth in the kHz range. The bandwidth (BW) will be used to focus or excite a range of spatial locations to generate a slice selection.

3) The slice gradient also contributes to the slice size. A larger gradient produces a smaller slice.

Both the transmitter and receiver operate in the RF frequency range. The RF transmit signal tips the spins (excitation). This transmission of energy occurs over a range of frequencies known as the excitation bandwidth (BW). To explore this concept further, we will start with a DLS to understand the principles of slice selection and how it relates to frequency. Let's compare the differences in the following two images shown in figure 5.3.

Not all of this may be intuitive at first. As you practice or rehearse with these images, it will provide you more capabilities to discuss ideas with vendors from whom you are purchasing products, read and understand publications in the area, and/or enhance your workflow in the clinic.

From the DLS, you might observe some of the following:
1. The bandwidth signal determines the slice thickness.
2. The duration of the RF pulse (in time) is inversely related to the bandwidth of the signal (i.e., to achieve a large frequency BW, the duration of the RF pulse is shortened), as portrayed in figure 5.4.
3. Astute observers may have noticed that there appears to be a relationship between the frequency response of the RF excitation and space. Slice selection is dependent on both the RF-transmit BW and the frequency gradient. The frequency gradient encodes spatial localization, while the RF-transmit BW uses this encoding to determine where the slice selection occurs (see chapter 2 to review spatial gradients to relate them to spatial localization). Altering the frequency gradient slope will alter the slice thickness. We will expand on this topic shortly.

Note that the energy is related to the square of the field strength, as we will discuss in chapter 6 (also recall the Larmor equation, $\omega = \gamma B_0$). You might now recognize

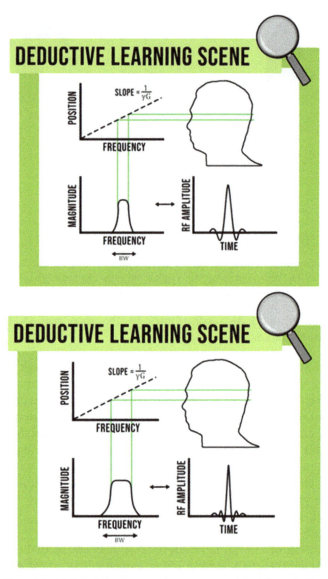

Figure 5.3. DLS depicting how varying parameters define slice selection.

that there is a linear connection in MRI between the frequency and magnetic field by design[4]. Thus, a square in B_0 corresponds to a square in frequency in MRI. You may recall from section 1.14 how we discussed the implication for increased wavelength

---

[4] There always can be exceptions to this case of linear gradients. Examples can be found in some advanced pulse sequence design, such as spiral and/or quadratic or other nonlinear gradients (more the domain of MRI pulse sequence designers), which are much beyond this scope of this book. Further reading can be done in the *Handbook of MRI Pulse Sequences* [1]. For clinical learning purposes, however, it is essential for readers to follow along with the key linear logic of a constant spatial gradient during readout and spatial position.

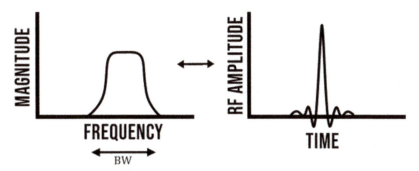

Figure 5.4. Duration of the RF pulse is inversely related to the frequency bandwidth.

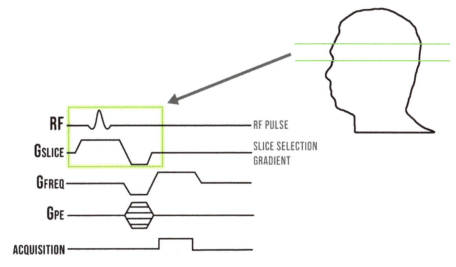

Figure 5.5. Looking back at the pulse sequence diagram, the two elements used for slice selection are the RF pulse (notice the shape) and the slice selection gradient (Gslice).

with energy. Next, we will review the appearance of the RF and slice select gradient directly on the pulse sequence diagram in chapter 4, shown in figure 5.5.

 Natalie N. (premed) asked: "I have heard of another type of bandwidth that has to do with frequency listening. Is this the same bandwidth you are talking about?"

Gail H. (neuroscience postdoctoral researcher) said "'Receive bandwidth' is describing the range of frequencies that you are 'listening to' during a scan. On the other hand, you also have 'transmit bandwidth,' which has to do with the RF pulse used to tip the spins during the scan. It's important to distinguish between the two types of bandwidth."

Dr. Wu also added: "On the scanner, we typically see the 'bandwidth' as a selectable parameter of the receiver (this isn't what is controlling the transmitter bandwidth). The bandwidth of the transmitter is buried in the SAR calculations and not usually visible to the technologists. Bandwidth can be a great point of confusion because there are two kinds (the receiver and transmitter). One has to do with the receiver, which we will talk about later, and the other is the transmit bandwidth, which we described in the above section. The transmit bandwidth is the set of RF frequencies that the RF transmitter sends out. It incorporates the energy slice selection and is combined with the gradient. You may recall spatial encoding when you hear the word gradient. The combination of the two factors actually ends up dictating the slice thickness, which you can calculate using formula 5.1."

Recipe for slice selection:
1. Apply RF pulse. Use the Fourier transformation to generate a frequency response.
2. From the frequency response, find the bandwidth (range of frequencies that are to be excited).
3. The BW is converted via a 'gradient' slope (remember frequencies and space are interrelated via formula (5.1)).
4. This size and location of the slice is dictated by the BW and gradient applied by the system.

Now that you have the recipe, can you use the following formula?

$$\text{ST} \propto \frac{\Delta BW}{\gamma G}(*) \qquad (5.1)$$

ST = slice thickness is proportional to transmit bandwidth (BW) divided by the gradient strength (G) and γ is the gyromagnetic ratio, which is a constant (has units of MHz/T).

Consider the following figure. When only the slope of the gradient is increased, while transmit BW and excitation pulse are kept constant, the effect of the gradient

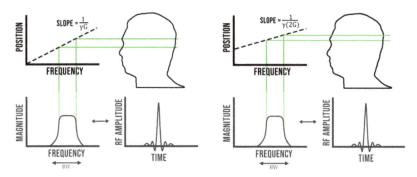

Figure 5.6. When only the gradient slope is doubled, the slice thickness decreases by half.

increases as well, resulting in a thinner slice. Another way to think of this is that it requires more 'energy'[5] to increase the gradient and the 'benefit' created is achieving a smaller slice or finer resolution in the slice direction, as portrayed in figure 5.6.

Applied bubble

If you would like some more practice, consider the following three thought exercises:
1. If we decrease the gradient strength to 1/3 of the original, what would happen to the slice thickness?
2. If we increase the transmit bandwidth by two, what would happen to the slice thickness?
3. If a vendor was able to provide you with higher gradients, then what would be the advantage?

Answers:
1. When gradient is decreased to 1/3, the slice thickness is 3 times as wide.
2. When BW is doubled, the slice thickness is doubled[6].
3. If a vendor was able to provide higher maximum gradients, then one of the advantages is that the vendor would be able to generate thinner slices.

---

[5] The author likes to consider the benefits of improved ability to deliver and/or consume energy to contemplate what potential advantages are achieved by that improvement. Here, an availability of more capable gradients (in a sense would require more energy) delivers higher resolution imaging.

[6] Note that decreasing the transmission bandwidth is the same as lengthening the time of delivery of the pulse with the same area to provide the same excitation. You can consider a decrease in transmission bandwidth to be reducing the energy. By decreasing the bandwidth, you would generate a wider slice (i.e., lower resolution), which would require less energy by the system than would attempting to generate a thinner slice.

Carlie P. (premed) reflects on how the transmitter, flip angle, and gradient relate to thickness: "I remember the transmitter excites different frequencies over a range, which is called bandwidth. If you excite a bigger range of spatial frequencies (bandwidth), then the slice thickness will also be bigger because it is directly proportional. Gradient strength also factors in to slice thickness. If you increase the slope of the gradient, the slice thickness gets smaller because it is inversely proportional. I can't remember if flip angle has to do with slice thickness so I'm going to ask Dr. Wu for some help."

Dr. Wu said: "This is a good question; the flip angle doesn't necessarily have to do with thickness, but rather the amount of tipping of the spins, which can affect the image contrast."

5.2 Traversing with *k*-space and qualities of the image

In daily conversation or perhaps at conferences, you may hear these phrases: SNR, contrast-to-noise (CNR), *k*-space, and spatial resolution.

K-space is the acquired spatial-frequency raw data. *K*-space may sound complex, but by understanding the 'wave' concepts discussed in chapter 1, you can better relate to the concept of *k*-space and the spectral domain. Remember that *k*-space is the space of mathematically added up 2D waves. This chapter aims to provide the reader with tools that can be used to enhance your understanding. Before we continue, consider the following: a single point in *k*-space does not directly correlate to a point within the image. Rather, each point of *k*-space provides some information applicable to every point of the image, shown in figure 5.7.

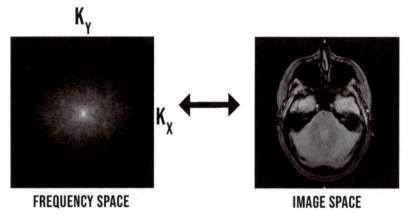

Figure 5.7. A scan of the author's brain, which has been acquired and transformed into frequency space, i.e., Fourier transformed.

Learning objectives:
1) The goal is to retrieve a spectrum from multiple dimensions (2D or 3D). For example, after slice selection in a 2D MRI sequence, it will have two spectral dimensions, i.e., frequency encoding and phase encoding directions, which we described in chapter 4 (see sections 4.32 and 4.42).
2) Areas of k-space vary in spatial-frequency. The center of k-space has a lower frequency, and contributes to the 'contrast' of the image, while the peripheral points are of higher frequency and contribute to the resolution. Sampling more in one area versus another will alter the appearance of the image.
3) The goal of a pulse sequence is to apply multiple gradients. The purpose of the gradients is to traverse the entire spectral space (k-space) and acquire the spectral data. The 'Fourier transformation' [2] is then applied to turn the spectral data into an image.

The understanding of spectral space and gradients provides a foundation for better understanding artifacts associated with the phase encoding versus frequency encoding direction. A stronger intuition for identifying artifacts and their causes allows you to more rapidly diagnose issues with image quality.

5.2.1 Retrieving spectrum in multiple dimensions and the center versus periphery k-space

It is instructive to take a second look at the square wave as we add the higher and higher y terms into the images. Recall that we 'previewed' this concept in chapter 1, when we previously introduced waves. Wave functions are sequentially added, as shown in figure 5.8, to construct the square function.

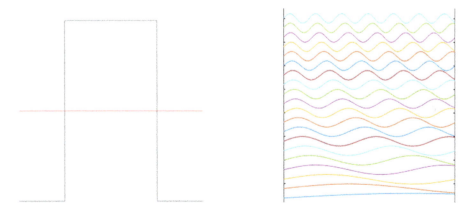

Figure 5.8. The low frequency waves provide the shape, while the high frequency waves show the edges. The figure to the left shows the addition of waves from lower to higher frequency added together that are meant to form the square wave. A small screenshot is shown in the figure to the right, which shows the different frequencies that are added together (note that they will have different amplitudes, which is not shown in this figure). Animation available at https://doi.org/10.1088/978-0-7503-1284-4.

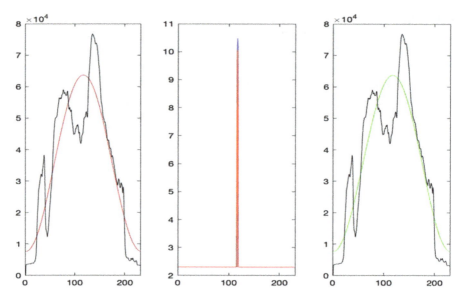

Figure 5.9. Consider the following animation, which is actually the project profile of the author's own brain with MRI. In (A) the illustration contains the full reconstructed image with the harmonics (shown in red) progressively moving away from the center (note the red profile in (B)), which is a 'zoomed' in version of the spectral components that would generate the corresponding signal. In (C), the sum of components is given in green matched against the original profile shown in black. Animation available at https://doi.org/10.1088/978-0-7503-1284-4.

Next, let us look at a more 'real life' example. Let's first take a single dimension (1D) image and look at the properties of the 'center' of k-space as we continue to add a greater and greater number of spectral components toward the outside of the figure, as shown in figure 5.9.

Dr. Justin N. (radiology attending) reflects on the k-space concept: "I just think of it as a bunch of energy acquired in another space where the image is encoded. It's hard to think of or visualize directly."

Whitney (MRI technologist) said about the k-space concept: "I think of k-space as just where the image is made."

Dr. Wu adds: "*k*-space can be initially a very complicated and abstract concept for many learners to consider. Perhaps one way to understand it is that it's simply the spectral space of frequencies that you are receiving from the sample. More intuition comes over time when you experience more and work with concepts, such as spectral changes that impact images. For example, in chapter 7, we will come across artifacts that are definitely impacted by an understanding of spectral frequencies."

Just like it is possible to create a one-dimensional Fourier representation of objects, it is also possible to do this in higher-order dimensions (i.e., 2D or even 3D). For an example, we use the famous self-portrait of Vincent Van Gogh and generate the *k*-space 'equivalent' for that image in the next figure, which is a two-dimensional image, as depicted in figure 5.10.

Figure 5.10. Van Gogh self-portrait original that has been inverse Fourier transformed in 2D to create image B. This 'File:Vincent van Gogh - Self-Portrait - Google Art Project.jpg' image has been obtained by the author from the Wikimedia website, where it is stated to have been released into the public domain. It is included within this article on that basis.

Recall that the points of *k*-space do not directly coordinate with points of the image. So how do we interpret *k*-space? The edges, or periphery, of *k*-space provide *low* spatial frequencies. Low spatial frequencies provide outlines, but not details. Consider the following image. If we set the center of *k*-space to zero, we are left with only the periphery values. Compare the resulting Fourier transform in C of figure 5.11 to the original above in figure 5.10. We can see the general outline, but lack details. This is because the center of *k*-space provides contrast.

What if we focus only on the center values of *k*-space and set the periphery values to zero? The result is an image that has contrast, but lacks the details (the resolution) and edges of the figure. The center *k*-spaces are the general carriers of the signal of the image, as illustrated in figure 5.12.

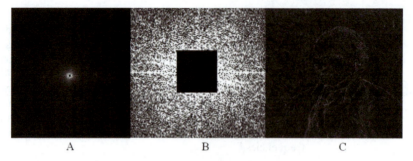

Figure 5.11. (A) The peripheral pixels of 3000 × 3000 are preserved; the center (20 × 20) is set to zero. (B) A close-up of the area around the center of k-space that is set to zero. (C) Resulting image without center k-space values; edges (resolution) are preserved, but contrast is lost.

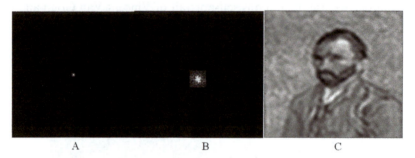

Figure 5.12. (A) The center pixels (20 × 20) are preserved, while the outer parts are set to zero. (B) A close-up of the preserved area around the center of k-space. (C) The resulting image provides the general shape and signal value levels of the original. However, the resolution is lost, resulting in the image appearing 'fuzzy.'

Nicholas P. (third-year med student) reflects on the difference between the inside and outside of k-spaces: "How can you demarcate the difference between the inside and the outside of the k-space?"

Dr. Wu added: "This is a critically testable topic. When you remove the outside of a k-space, you get a blurry image. However, that image contains most of the information carriers that define the contrast of the image. If you remove the center of the image, which can be quite small, you get the edges of that image without the main signal."

5.2.2 Traversing *k*-space with gradient structure of the pulse sequence diagram

Every point of *k*-space provides a more detailed image. **The goal of a pulse sequence is to acquire the entire *k*-space.** The raw data acquired can then be transformed into an image using the Fourier transform. The following animation provides a schematic of how we use phase encoding and acquisition intervals to obtain all *k*-space data. In reality, many more lines are used to minimize all gaps, animated in figure 5.13.

There are various 'patterns,' or trajectories, for *k*-space coverage. As long as every data point is acquired, the order of retrieval would modify the overall contrast of the image. The previous animation uses Cartesian coverage. Cartesian coverage (i.e., rectilinear) uses line by line traversal of *k*-space and is very common in MRI. The previous examples of pulse sequences display Cartesian coverage, as it is the simplest to understand, as shown in figure 5.14.

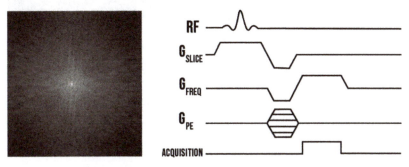

Figure 5.13. *K*-space is traversed 'step by step' using both phase encoding and acquisition intervals. Animation available at https://doi.org/10.1088/978-0-7503-1284-4.

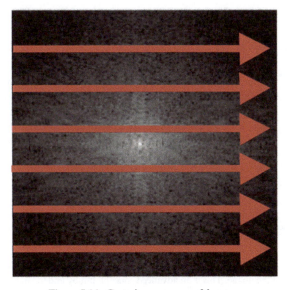

Figure 5.14. Cartesian coverage of *k*-space.

5.2.3 Scaffolding bubble

Advanced patterns for traversing pulse sequence

These gradients are 'played out',[7] along with RF pulses by the pulse sequences, to create the images in MRI. Each pattern has different strengths and weaknesses, but each has an elegant pattern, as shown in figure 5.15.

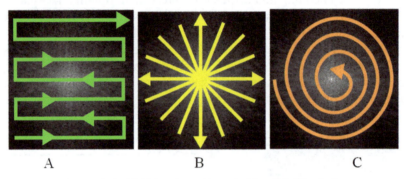

Figure 5.15. (A) Echo-planer (zig-zag), (B) radial, (C) spiral.

The radial pattern covers the center many more times than the outside. The reader might contemplate what the consequences would be.[8]

A) The first is a zig-zag pattern commonly associated with echo planar imaging (EPI) coverage [3].
B) The second is a radial pattern, which is associated with the propeller or radial family of pulse sequences [4].

 Gail H. (neuroscience postdoctoral researcher) reflected on why would we want to have different coverages of the k-space and how do the pulse sequence 'gradients' traverse k-space:

---

[7] 'Played out' is parlance used by MRI physicists/scientists to describe the timing and amplitudes of how the multiple gradients are turned on/off in a pulse sequence in the frequency, phase, and slice directions (figure 5.15).

[8] Covering the center more means more contribution to the 'contrast' of the image, perhaps at the expense of some spatial resolution in the image. This is an advanced concept, but by understanding the basic concepts you will continue to develop greater intuition.

"The way gradients can transverse k-space is if they focus on a certain area it can focus on things like contrast or edges, depending on how much you spend time in that part of the image. I am curious about what happens when you cover the same spot in k-space more than once. For example, in the radial pattern, the middle gets covered repeatedly. Do the values average? How does that work? Also, how exactly is this repeated sampling useful for image quality? I think that covering the middle versus the outside might more affect main signal vs. contrast/edges, but how does it work?

To answer these questions, I went to 'MRIquestions.com.' Apparently, oversampling certain areas can be helpful because you can compare data between passes and correct for motion artifact (radial pattern see [5]). The duplicate values for the same spot can be thrown out if there is an issue. Also, because k-space gets more thoroughly 'filled-out' in the middle in a radial pattern, there is greater emphasis on improved signal-to-noise ratio. Since the edges of k-space get covered less, there is less information entered there, but it isn't as important to have it all filled out for the bulk of the signal to come through."

Dr. Wu added: "To answer the first question, there are many reasons to have different traversal patterns. First of all, there are different image weightings, and the traversal patterns will help dictate those weightings. Furthermore, there are some advantages with signal-to-noise accentuation of things, like contrast over resolution and a bevy of different types of artifact control."

5.3 On the receiver side

The third portion that is covered in this chapter concerns the reception of the MRI signal (figure 5.16). We covered the ideas of 'contrast' in chapter 3 when we described relaxation. Now, to advance these concepts, we will discuss the notions of signal and contrast in the presence of noise. We will next work our way through the concepts of signal-to-noise in section 5.3.1, then contrast-to-noise in section 5.3.2 followed by the critical topic of receiver bandwidth in section 5.3.3 of this chapter.

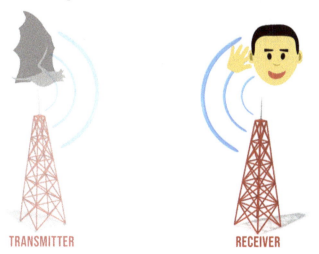

Figure 5.16. Focus on the reception portion of this figure for this section.

Figure 5.17. An image of a city with significant haze. As there is more haze, it becomes more difficult to be sure that what you are seeing in the background is what you believe it to be.

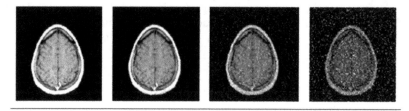

Figure 5.18. From a strict SNR definition, it is the ratio of the average signal value μ to the standard deviation of the signal σ. Note in the figure that from left to right additional 'noise' was present in the images, decreasing the 'confidence' of being able to read the structures through the 'haze' of that noise.

5.3.1 Signal-to-noise ratio (SNR)

It is imperative to distinguish SNR in order to produce a quality image. **SNR can be generally thought of as the confidence that you have in your images**. High SNR means that you can observe your signal above the haze (noise), as portrayed in figure 5.17. 'Noise' will result in the image appearing grainy or containing irregular patterns. The noise is a result of molecular movements within the body and electrical resistance within the MRI machine itself.

For technologists, acquiring sufficient SNR acts as a quality assurance measure. However, contrast-to-noise may be even more valuable, but is not always readily apparent without the input of the radiologist regarding where to look, as shown in figure 5.18.

$$\text{SNR} = \frac{\mu_{\text{sig}}}{\sigma_{\text{sig}}} \tag{5.2}$$

where μ_{sig} is the mean of the signal and σ_{sig} is the standard deviation of the signal.

5.3.2 Contrast-to-noise

Contrast-to-noise ratio (CNR) follows the same guidance for SNR. Recall that contrast is the ability to distinguish. **CNR regards the confidence in distinguishing between tissue types, including healthy versus diseased tissues.** Detecting differences between diseased tissues and normal tissues is of particular importance. An image with high amounts of noise will reduce the contrast of the image and may mask an abnormality.

$$C = \frac{|S_A - S_B|}{\sigma_o} \quad (5.3)$$

Where S_A and S_B are signal intensities for signal producing structures A and B, and σ_o is the standard deviation, as measured in two regions of interest.

5.3.3 Bandwidth (receiver): a listening perspective

If you still feel unsure of thinking in the frequency domain, nobody can blame you, as it takes time and practice to gain mastery (figure 5.19). The trick is to conceptualize 'sampling time' first, as it may be more intuitive to understand that sampling longer yields more SNR. Bandwidth is inversely proportional to sampling time; thus, a higher BW will warrant a shorter sampling time and yield a smaller SNR. Remember that your mastery of MRI can depend on you continuing to strengthen your intuition on thinking in both the space domain (spatial frequency) and in the time domain (frequency in terms of time). The ability to mentally process both arguments in both domains can be a benefit and help you with the materials in the later chapters of this book, as well as your clinical service.

Similar to a window, the wider you 'open' your bandwidth, the more light or noise you will allow through, as portrayed in figure 5.20. When the window is barely open,

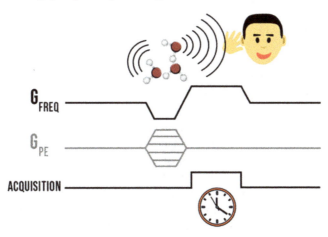

Figure 5.19. When you first think of bandwidth, think of listening to a signal. The spins emit signal at varied frequencies. BW determines which frequencies we 'tune' into. You measure each sample for a given time (called the sampling time).

Figure 5.20. Bandwidth is inversely related to the sampling time. As you listen longer, typically you have more ability to discern the signal, providing a greater SNR. This 'File:The open window (6028681236).jpg' image [7] has been obtained by the author from the Wikimedia website where it was made available marcsiegert under a CC BY 2.0 licence. It is included within this article on that basis. It is attributed to marcsiegert.

Figure 5.21. Bandwidth can be viewed as a window. You choose how wide to open the window.

you see a clear, strong beam of light. The wider the window is opened, the less defined the beam of light becomes. Thus, selecting a wider, or larger, receiver bandwidth presents greater noise and a worse SNR. However, small bandwidths tend to cause chemical shift artifacts.

The MR signal is based at an RF frequency, which we will see later is fixed by the strength of the main magnetic field, but it contains many different frequencies that encode information about the location of various tissues (figure 5.21) [8]. The base RF frequency is removed from the signal before it is digitized, leaving the receiver bandwidth of the signal, which is typically several kilohertz wide. Most of the signal-to-noise and contrast information is at very low frequencies, while the higher frequencies contain information about resolution in the image (see 'An easy introduction to k-space'). Electronic noise is distributed evenly across the whole bandwidth, as depicted in figure 5.22.

Understanding bandwidth is important, as it can be optimized for a variety of possibilities, and also because changing the bandwidth affects the timing between RF excitations of the system, and can affect motion/movement control in the acquisition, sensitivity to chemical shift artifacts (see chapter 7), and contrast in the image (through affecting the image via the relaxation mechanisms).

Natalie's reflection: How was this graph helpful in clarifying how bandwidth affects the timing between excitations and the subsequent effect on resolution? "This graph helped me realize how bandwidth affects timing because, as shown in the image, when there is a smaller window, it allows for more SNR, which creates a clearer image as shown above. I learned that bandwidth and sampling time are inversely related, so when you have a high bandwidth, you will have a smaller sampling time, which will yield a lower SNR, which, as shown above, fills the images with more noise. It's helpful to have the window analogy to think directly in terms of what is happening in the frequency space, i.e., when the window (or bandwidth) is wider, it allows more noise to be let in, but the sunlight comes in, regardless."

Yield bubble: oversampling

The purpose of oversampling is to reduce wrap-around and aliasing artifacts by increasing the field of view (FOV).
- Oversampling in the frequency direction acts to suppress aliasing artifacts. The acquisition runs at double the sampling rate and acquires double the amount of samples. Most systems automatically apply frequency oversampling. **Frequency oversampling does not alter the SNR.**
- Phase oversampling is similar in principle. By applying more phase-encoding steps, we can increase the FOV to avoid wrap-around artifacts. Every step that is added will **increase the acquisition time**. Therefore, phase oversampling is not automatically applied and must be controlled by the operator. In general, **phase oversampling will not change the SNR**.

We will revisit this topic in chapter 6 for tradeoffs and in chapter 7 for the aliasing artifact. More detailed information can be found in [6].

These elements will help you better master tradeoffs in the next chapter, as well as in chapter 7, which will cover artifacts. With these items in your repertoire, as displayed in figure 5.23, you are hopefully well fortified to proceed to SNR and time tradeoffs, which are important to optimize sequences.

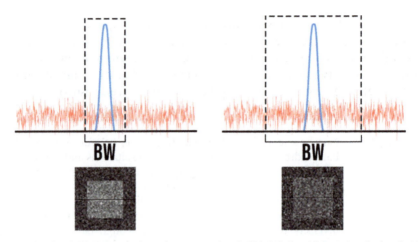

Figure 5.22. The signal (blue) travels through a narrow bandwidth 'window.' Increasing the bandwidth risks letting in too much noise (red), resulting in a poor SNR. Note, the signal comes in over a small range of frequencies (i.e., comes small bandwidth).

Figure 5.23. If you have mastered this section, you now have some of the bread-and-butter elements (Adobe Stock Images: © sriba3). On the right, we show the three items that you have mastered by completing this chapter.

5.4 Summary

Now you are ready to get serious when it comes to MRI. We can next try to apply these new skills to do some challenging, yet practical concepts, like tradeoffs and artifacts, and explore some of what you may encounter day-to-day in your practice. The fundamental idea to master is relaxation in chapter 3, because it is the manipulation of parameters that accentuate the contrast in terms of T1 and T2. To give you a chance to master them in daily conversation, you might hear the phrases signal-to-noise, *k*-space, bandwidth, and/or spatial resolution.

It was important to realize that the three ideas are interrelated and therefore require tradeoffs to acquire an ideal image. The following concepts, as well as the parameters that control them, will provide a foundation when we begin our discussion on artifacts. Tradeoffs in MRI are further described in chapter 6. MRI artifacts are further described in chapter 7.

Recall the discussion in chapter 4 on pulse sequences, particularly RF tipping with slice selection. We provided further details on that concept in this chapter in regards to slice selection. For the RF excitation pulse, it is important to understand how the transmit bandwidth works. This will also work in conjunction with the slice gradient, as we described in chapter 4 section 4.3.1.

Another topic concerns the spatial gradients and k-space. Building off the understanding of k-space and Fourier transform, we will proceed and go into more detail on navigating k-space. Specifically, we compared the impacts that result from changes to the inside and outside of k-space. Each will have a separate impact on image resolution and/or image contrast.

The third concept of this chapter revolved around the signal received by the RF receiver, particularly the SNR. This requires an understanding of bandwidth, briefly discussed in chapter 4. We will now discuss how the receiver bandwidth size influences the SNR.

Some of this may feel abstract and maybe initially a little hard to reach, but many attendings comprehend these ideas, and it is important to have exposure to them if you want to optimize your practice[9]. We walked through three major bread-and-butter parts of MRI that will help you with understanding tradeoffs and artifacts, which is the subject of the next two chapters.

This chapter primarily addressed the following ideas:
1) We stepped back and revisited tipping as provided by the RF transmitter. Two lines of the pulse sequence (i.e., RF and G_{slice}) were illustrated in the previous chapter. However, we connected the 'gradient' and bandwidth of tipping together with spatial localization through slice selection in the first section of this chapter.
2) Then, we discussed the 'dreaded by some' K-space[10]. 'Rules of k-space: (a) the goal of pulse sequences is to 'cover' it and (b) where we cover impacts the image (center and outside of k-space).
3) Next, we covered noise and seeing through it—signal-to-noise ratio = confidence you can have in separating objects and disease from non-disease in the presence of noise.
4) Last, we looked at concepts surrounding receiver bandwidth and sampling time, which can impact your signal-to-noise in your acquired images.

---

[9] Researchers will also benefit from this knowledge, as it is practical and what is used to form decisions not only in the clinic, but also in experimental design. Intuition about the MRI clinical practice is most useful when translational research projects are planned or conducted.

[10] In chapter 1, we discussed acquisition through waves and spectrum. If you were able to intuit through the activities the idea of how object/image shapes can be synthesized as a wave and the Fourier transform, perhaps this can remove some of the 'dread' that comes with this technical-sounding term.

References

[1] Bernstein M A, King M F and Zhou X J 2004 *Handbook of MRI Pulse Sequences* (Oxford: Elsevier) pp 267–97
[2] Fourier transform (Radiology Reference Article, Radiopaedia.org (Retrieved 26 January 2022) https://radiopaedia.org/articles/fourier-transform
[3] Stehling M K, Turner R and Mansfield P 1991 Echo-planar imaging: magnetic resonance imaging in a fraction of a second *Science* **254** 43–50
[4] Pipe J G 1999 Motion correction with PROPELLER MRI: Application to head motion and free-breathing cardiac imaging *Magn. Res. Med.* **42** 963–9
[5] Radial sampling (n.d.). Questions and Answers in MRI (Retrieved April 6, 2022) http://mriquestions.com/radial-sampling.html
[6] Brown R W, Cheng Y-C N, Haacke E M, Thompson M R and Venkatesan R 2014 *Magnetic Resonance Imaging: Physical Principles and Sequence Design* 2nd edn (New York: Wiley)
[7] https://en.wikiquote.org/wiki/File:The_open_window_(6028681236).jpg
[8] McRobbie D W, Moore E A, Graves M J and Prince M R 2017 *MRI from Picture to Proton* (Cambridge: Cambridge University Press)

IOP Publishing

MRI: Connecting the Dots
A start to concepts
Dee Wu

Chapter 6

Three tradeoffs in MRI (Clinically relevant)

(Balancing SNR, balancing time, balancing heating)

6.1 Introduction

In the previous chapters, we looked at a pulse sequence line by line and discussed the corresponding parameters that control image quality. In reality, these parameters do not act independently, but require tradeoffs between each. Changing one will affect the others. We will focus on three qualities: SNR, resolution, and acquisition time. Selecting the correct pulse sequence requires an understanding of how to balance parameters in order to attain the highest image quality. Taking into consideration the body part being scanned, the circumstances of the patient, and the suspected abnormality, the radiologist and radiographer have the important job of selecting an effective protocol and monitoring the scan, as illustrated in figure 6.1.

As the saying goes, 'you can do things fast, cheap, or good.' Let's tweak this a bit for MRI:
- Fast—obtaining more samples increases the duration of the scan, but can improve signal.
- Good—SNR is a crucial indicator of image quality.
- Cheap—increasing exposure to RF energy may come with a cost to patient safety. Specific absorption ratio (SAR) measures the effect of RF energy on a patient.

6.2 Factors that change time

This section provides guidance on how to approximate the time for how long a sequence would take (figure 6.2).
  Learning objectives:
  1. Understand the impact of TR on the time of a sequence.
  2. Understand the impact of increasing the number of averages on SNR.
  3. Understand the impact of increasing the number of PE on the SNR.

Figure 6.1. Adjusting image parameters requires balancing tradeoffs between time, SNR, and specific absorption ratio (SAR). A balance between the three is essential to producing a quality image.

Figure 6.2. Time as a critical factor to manage in the clinic and pulse sequences. Reproduced with permission from Quality Stock Arts/stock.adobe.com.

A general formula to use for time is[1]:

$$\text{Sequence time} = N_{\text{AVG}} * N_{\text{PE}} * \text{TR} \quad (6.1)$$

---

[1] Note the formula is for the more basic sequences. However, the reader may come across some sequences that don't fully obey this basic formula. For example, if the sequence is a fast spin echo (FSE), then the total sequence time equation will be modified to be sequence time = $(N_{\text{AVG}} * N_{\text{PE}} * \text{TR})/\text{ETL}$, where ETL is the echo train length of that FSE sequence. There are even more sophisticated patterns of time and SNR estimates for pulse sequences that can include compromises from parallel imaging as well as up and coming denoising techniques such as deep recon and compressed sensing which are beyond the scope of this book, but we mention here for future reference these advanced and emerging concepts.

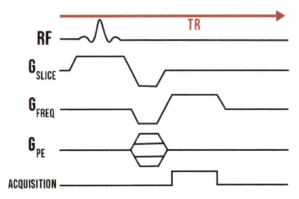

Figure 6.3. Pictorial reminder of the repetition time (TR) in a pulse sequence diagram (PSD). Note that each time the sequence block of the PSD is repeated, the time will go up by the TR. Animation available at https://doi.org/10.1088/978-0-7503-1284-4.

Now we are going to try to break this down by parts. We will examine what happens to time when changing repetition time (TR) as seen in figure 6.3, and by changing the number of averages $N_{AVG)}$ and/or number of phase encode steps (N_{PE}). We will break these concepts down in the next few subsections and boxes 6.21 and 6.22.

6.2.1 Repetition time (TR)

Repetition time (TR) is the amount of time between successive RF pulses. A pulse sequence block is the repeated section of the pulse sequence and lasts the duration of one TR, as shown in figure 6.3[2]. It can be determined by the time between the excitation RF pulse of the first iteration and the repeating same excitation RF pulse of the next iteration. Note that the TR has an important effect on the control of image contrast characteristics, particularly in regard to the image weighting. The effects on image contrast affecting T1 and T2 weighting are described in chapter 3 in detail.

Examples of how TR can change time:
How does TR change the total time of a sequence?
A: We needed to save a few seconds on the time of the sequence, so we lowered the TR from 600 to 400 for a T1 scan.

---

[2] Repetition time (TR) was described in section 4.3

> B. We needed to lower the specific absorption ratio (SAR)[3] of a sequence, so we increased the TR, but the tradeoff was increasing the time of the scan.

6.2.2 Number of phase-encoding steps and number of averages effects change on time

> As described previously, blocks in figure 6.3 are repeated. There are two forms of repetition of blocks with the number of averages (N_{AVG}) and number of phase encoding (N_{PE}) blocks.

First, let us focus on N_{PE}, the number of phase-encoding steps. We will first assume that there is only one average ($N_{AVG} = 1$)[4]. As an example, for figure 6.4, this would be $N_{PE} = 6$ repeats of the pulse sequence block each, which would take TR time per block. While some of this material may take a little practice, some practice will provide the reader with some basic conceptual knowledge of how number of phase-encoding steps and number of averages affect the overall duration of the sequence. The knowledge about the duration of scans is helpful for technologists and radiologists so that they can manage the efficiency of their scanners.

If we dissect the 'toy' pulse sequence diagram above, we have six PE steps ($N_{PE} = 6$). The pulse sequence block must then be repeated a total of four times, for a total of 6*TR (e.g., if TR = 500 ms, then total time is $T = 6*500$ ms, or 3s). The overall duration of the

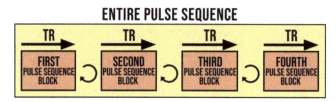

Figure 6.4. Repetition time is multiplied every time that block is repeated. The above example illustrates a specific number of $N_{AVG} * N_{PE} = 6$. This is a small number and is just used for illustration purposes.

---

[3] We preview the specific absorption ratio (SAR) at this point so that the reader can prepare to integrate the concept. SAR concerns heating/safety effects in MRI and will be covered in section 6.3 of this chapter.
[4] Averaging is a method that is used to improve the signal-to-noise. The effect on signal-to-noise ratio (SNR) will be covered in more detail in section 6.3.1–6.3.5 later in this chapter.

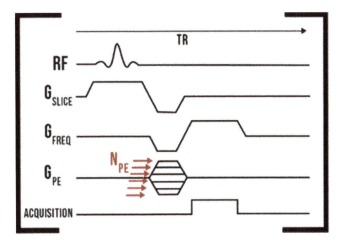

Figure 6.5. Repetition time (TR) is multiplied every time that block is repeated. The above example illustrates a specific number of $N_{PE} = 6$. This small number is used for illustration purposes in the figure to reduce clutter in the number of lines.

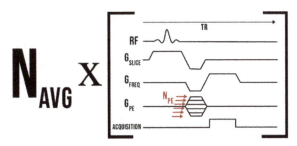

Figure 6.6. Number of Averages (N_{AVG}) refers to each time an entire pulse sequence is repeated. Number of Pulse Sequences (N_{PE}) each take into account a repeated block of acquisition. Thus, the total number of times that the acquisition block is acquired is N_{AVG} times N_{PE}. Note that this figure is similar to figure 6.5, with the addition of the second loop over TR that involves Number of Averages (N_{AVG}).

scan is dependent on a variety of factors, including averages, number of phase-encoding steps[5], and TR, which influence the total acquisition time, as shown in figure 6.5.

When you are performing averages, then you are looping over the sequence block, as shown in figure 6.6. This sequence block includes multiple phase-encoding loops (NPE) that are applied over a TR.

---

[5] The scanner value for N_{PE} phase-encoding steps is 192, for example. Then, a more realistic sample for $N_{AVG} = 1$, so 192 × 500 ms *(1s/1000 ms) = 96 s, or a 1 min and 36 s scan time for just one average. As we will soon see, for a $N_{AVG} = 2$ this scan may approach 3 min, which may be roughly around the values we would see around a sagittal T1-weighted spine MRI scan.

Whitney says: "Managing time of sequences is something I do a lot in my daily work as an MRI tech. This is not only to make our scanner more efficient, but also to help with patient compliance. Since some patients have health challenges, it is hard for them to stay still as time of scans increases."

Managing the Averages (N_{avr}):
"One way I do this is that I manage the number of averages. These number of averages can double time if ($N_{avr} = 1$ to $N_{avr} = 2$), or increase time by 1.5 from ($N_{avr} = 2$ to $N_{avr} = 3$)."

Managing the TR:
"There are other ways that time gets altered, like if I just change the TR sequence as long as I retain the correct weighting."

Managing the Number of Phase Encode Steps (N_{pe}): "Sometimes I encode the 256 to 192, so the time got shortened by 25% (192/256 = 0.75) just by reducing the N_{pe}."

6.2.3 Knowledge check

Typically, if there is a question (on boards, for example), it will be a relative measurement of changes in pulse sequence per TR. For example:
1. What happens if you double the number of averages (N_{AVG})?
2. What happens if you decrease the number of PE steps?

Possible answers
1. Doubling the number of averages increases time by a factor of two, and increases signal-to-noise by the square root of two.
2. Decreasing the number of PE steps decreases time and decreases signal-to-noise.

Scaffolding Bubble

A question that the astute reader may be interested in is why increasing the number of frequency steps does not immediately change the time of acquisition, while N_{pe} does (figure 6.7). Phase-encoding steps cause a pulse sequence to repeat, whereas frequency-encoding steps are built into the TR and do not increase the duration of the scan. Notice how the frequency encoding subdivides the acquisition window.

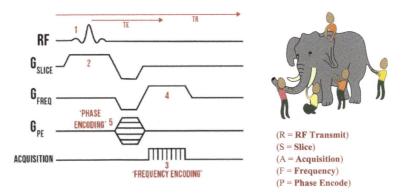

(R = RF Transmit)
(S = Slice)
(A = Acquisition)
(F = Frequency)
(P = Phase Encode)

Figure 6.7. Phase-encoding steps cause a pulse sequence to repeat, whereas frequency-encoding steps are built into the TR and do not increase the duration of the scan. Notice how the frequency encoding subdivides the acquisition window.

From Figure 6.7, it can be seen that the subdivision that the frequency-encoding steps incur affects the TR either not at all or very slightly[6]. However, you may note that for each phase encode that is conducted, there is another line that requires TR time. For visualization of phase encoding, the reader can examine the animation shown in figure 4.42 of appendix 4.A.3 in chapter 4.

6.3 SNR tradeoffs

While treatment of SNR may appear technical, signal-to-noise is important for the perception and reading of MRI images, as depicted in figure 6.8. As described in section 5.3.1 of chapter 5, **SNR can be generally thought of as the confidence that you have in your images**. Technologists devote much effort to managing SNR in images and to meeting the expectations of the readers and interpreters of those images[7].

Learning objectives for SNR
  1. Spatial size and abundance-related.

---

[6] It may affect it slightly if the acquisition window needs to be increased. However, this would only add a slight percentage to the TR, which would not necessarily change the time much. For a more detailed read, the reader is encouraged to consult [1] and/or [2]. Both books are excellent references.

[7] A hot topic even during the time of publication of this book in 2022 is specialized early artificial intelligence techniques that are used to de-noise the images. All major MRI manufacturers have versions of AI and reconstruction which have been released as specialized packages that have obtained FDA clearance. These technologies are exciting developments; however, each site must carefully consider the adoption dependent on their needs. Important discussions by the healthcare team as to which are acceptable modifications and as to which parts of the body, they are useful for regarding new AI techniques and balancing the concerns of generating new types of reconstruction artifacts, impact image contrast, and/or impact spatial resolution continues to need to be addressed. Image quality in these new technologies is still under investigation [3] and optimal levels of de-noising may need to be further assessed. Regardless, the management that balances image quality in MRI along with time efficiency are key concerns in a busy MRI facility and hospital environment.

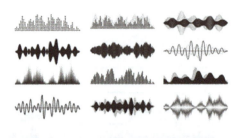
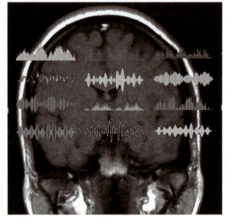

Figure 6.8. Collage signal-to-noise imagined. Note that signal-to-noise of scans is a major concern and important to manage in the clinic, as SNR relates to the 'confidence' a reader may have in their scans. Reproduced with permission from pixels/stock.adobe.com.

2. Averaging-related.
3. Time-related (review receiver bandwidth).
4. Others (coils, flip angle, crosstalk interference).

 If you see **formulas** below, **DON'T WORRY** if you are not used to applying them directly. The formulas are provided for convenience and summarization, and we have found that some residents find the structure of equations/formulas easier to memorize and apply. Other learners may utilize and synthesize the information better through tables[8].

---

[8] A couple of readers have suggested that when they see 'formulas,' it brings back some bad memories and they are worried about these formulas. We relate to your concerns, as any concept you don't see frequently will be a little nerve-wracking. But please, don't be overly concerned! For example, equation (6.1) is simply repeated several times as the same equation in sections 6.2–6.5. It just looks messy, but it is the same equation that has different red highlights as we dissect it for the reader to break it down. Also, online or scanner calculators serve a major role in making this easier for technologists at the workstation, and the author has worked extensively to make 'calculators' for the scanners that help simplify the job of the users, as discussed in section 6.5. Note, we will leave a picture of a calculator to remind the reader there are online tools that can help them with these formulae.

Whether you use formulas or table concepts, a sense of what happens when you change SNR parameters is an important part of MRI, and these impacts of parameter changes are board-testable concepts. Also important is that signal-to-noise management is a major concern and important to optimize performance in the clinic. What guidance do we have when selecting parameters? Luckily, there is a reasonably straightforward formula for sequence parameters (which you can also program into a calculator):

$$\text{SNR} \propto \text{FOV}_{RD}/\text{N}_{RD} * \text{FOV}_{PE}/\text{N}_{PE} * \Delta z * \sqrt{N_{AVG} * N_{PE}/BW_{RECEIVE}}$$

Where N_{AVG} is the number of averages of the system, N_{PE} is the number of phase-encoding steps, FOV_{PE} is the field of view in the phase-encoding direction, FOV_{RD} is the field of view in the read direction, and N_{RD} is the matrix in the read direction only[9]. Δz is the slice thickness.

According to 'cupcakejones.net,' How Much Sugar Is in a Slice of Cake? There were 36 grams of sugar per cake, on average 6±7 g per slice. The amount of water consumed is 6 g/100 g.

Analogy of sugar inside a slice of a brownie/cake with the amount of hydrogen protons in each division or voxel size (brownie image Adobe Stock Images: © Vladislav Nosik). As the volume size increases, the amount in population of spins increases, which would be the amount of signal-to-noise.

For general purposes, the scanner interface will provide the value for the FOV and the matrix. The FOV is the entire area being imaged, measured in distance. The matrix divides the FOV into voxels. The **matrix** is constructed from the frequency-encoding steps in one direction and the phase-encoding steps in the other direction, resulting in a grid-like structure. The matrix value is a number (for example, 320 × 320). Recall from chapter 4 that the pulse sequence directions read, phase, and slice. Note that any of the read, phase, and/or slice can be assigned to any direction x, y, z, or any vectorial direction.

---

[9] You can choose any direction (x, y, or z) to be the read, phase, or slice direction (provided there is only one choice).

6.3.1 Spatial size and abundance

 Let us think of SNR as a piece of 'cake.' In this case, the number of available hydrogen protons in a voxel is like the amount of 'sugar' in a slice of dessert, as depicted in figure 6.9. The more protons, the better! The FOV is equivalent to the size of the whole cake. The slicing of the cake is equivalent to the matrix.

Figure 6.9. SNR with piece of 'cake' illustration. The number of available hydrogen protons in a voxel is like the amount of 'sugar' in a slice of dessert.

A bigger cake means bigger pieces. A larger voxel will produce more signal. However, if we need more 'pieces,' we can increase the matrix, but the voxels will be smaller and the signal will decrease. Keep in mind that the voxels within an MRI scan are three-dimensional. The **matrix** determines the 'length and width,' but it is the slice selection that determines the depth. A thicker slice will increase the overall area of the voxel, thus increasing the number of 'spins' included. The more 'spins,' the stronger the signal and the better your SNR will be.

Recall chapter 5, slice **'thickness'** is directly correlated with transmit bandwidth—not to be confused with receiver bandwidth. Increasing the transmit bandwidth increases the slice selection. In addition to manipulating various parameters, increasing the strength of the overall magnetic field will also increase the signal.

Noise is always present in every voxel of an image. Signal is dependent on the presence of protons. The higher the number of protons present within a voxel, the stronger the signal will be. Therefore, voxels with higher proton density will deliver a better SNR. Note that noise is primarily a side effect from the electronics, as opposed to the subject.

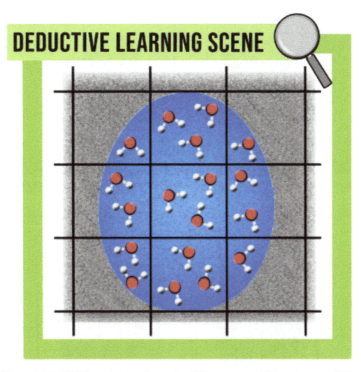

Figure 6.10. DLS investigating the trade-off between spatial location and SNR.

Using our equation for signal-to-noise relationship, refer to the DLS below to evaluate the relationship between FOV and abundance of protons, as well as SNR. We need only focus on the red parts of the equation below for the moment.

$$SNR \propto \frac{FOV_{RD}}{N_{RD}} \times \frac{FOV_{PE}}{N_{PE}} \times \Delta z \times \sqrt{N_{AVG} \times \frac{N_{PE}}{BW_{RECEIVE}}} \qquad (6.2)$$

Let the blue oval represent your object of interest. Because noise is primarily produced from the electronics, it remains constant, regardless of parameters.

Consider two possibilities. Focus on how the number of resonant molecules, or 'spins,' are altered, as depicted in figure 6.10.
- The matrix is constant. If you increase only the FOV, what happens to the SNR?
- What effect does increasing the matrix have on SNR?

Increasing the FOV will increase the number of protons present within a voxel, thus increasing the SNR change. Figure 6.11 illustrates the impact on the image

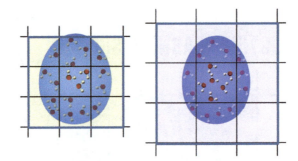

Figure 6.11. Focusing on the center voxel, the original FOV encompasses three protons. Increasing the FOV increases the size of each voxel. Note that the yellow area FOV of left and purple area FOV of right images are different. The enlarged voxel now encompasses six protons and would then have more overall signal (per/voxel), which translates to greater perceived signal across the entire image when examined by the viewer.

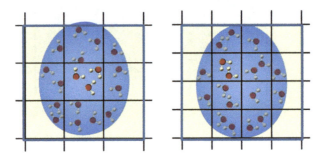

Figure 6.12. Increasing the matrix from (3×3) to (4×4) decreases the voxel size and the number of protons present per voxel. The result is a decrease in SNR. Note the yellow area would be maintained at the same FOV between the left and right images.

SNR per voxel when the overall FOV has been increased. While increasing the FOV may increase the signal per voxel, remember that this chapter is focusing on trade-offs. Increasing the FOV too much can have negative effects on spatial resolution and produce blurry images.

So what happens when we increase only the matrix? The voxels are decreased in size, resulting in fewer protons per voxel. Fewer signal carriers per voxel decreases the SNR. Selecting a **matrix size** that is too large will result in a grainy image. However, if the matrix is too large, the image will appear blurry due to poor spatial resolution, as illustrated in figure 6.12.

Let's tie this back into our equation for SNR.

$$SNR \propto \frac{FOV_{RD}}{N_{RD}} \times \frac{FOV_{PE}}{N_{PE}} \times \Delta z \times \sqrt{N_{AVG} \times \frac{N_{PE}}{BW_{RECEIVE}}} \quad (6.3)$$

Increasing the FOV — Increases the SNR linearly
Increasing the slice thickness (Δz) — Increases the SNR linearly

Whitney (MR technologist) said: "Sometimes I have to increase the field of view (FOV) from 14 cm–16 cm or even 18 cm FOV … picking the right FOV is a balancing act and we try to match the preferences of the radiologists. As technologists, we try to get all the anatomy of interest into the field view as what we think the radiologists need. Sometimes, technologists will need to be aware that increasing or decreasing the FOV can cause an effect on spatial resolution and SNR. A good example is that decreasing the FOV on a finger or toe, since they are some of the smallest body parts we scan, can lower your SNR and make it hard for our radiologists to read. This is a balancing act that we even have to manage on the fly."

Nick P.'s (medical student) reflection: "I have worked with radiologists who frequently call the technologists when they have any questions about the scans that they are reading. The technologists are very helpful in clarifying any confusion that may arise since they are the ones who obtained the imaging and know how they obtained the images. I can see how working closely with them in the future will be very helpful to work effectively as a radiologist."

Knowledge check:
- Understand how and why the FOV changes the SNR.
- Understand how and why the matrix changes the SNR.

6.3.2 Averaging

Averaging is another method to improve the SNR. Signal averaging relies on the fact that the measured signal is identical every time a scan is conducted, but the noise is random and therefore cancels out when multiple acquisitions are averaged. The signal, however, will accumulate and produce a better SNR.

Now let's focus on the next two terms inside of the square root:

$$SNR \propto \frac{FOV_{RD}}{N_{RD}} \times \frac{FOV_{PE}}{N_{PE}} \times \Delta z \times \sqrt{N_{AVG} \times \frac{N_{PE}}{BW_{RECEIVE}}} \quad (6.4)$$

Increasing the number of phase encodes (N_{AVG} or N_{PE})
Increases the SNR by the sqrt(N_{AVG}) or sqrt(N_{PE})

6.3.3 Scaffolding bubble: signal averaging

More formally, the noise is commonly expressed as the 'standard deviation of the mean,' which is typically N times smaller than the standard deviation of a single measurement. We model the averaging process by assuming a signal value $S(t)$, which is contaminated with a noise $N(t)$, as illustrated in figure 6.13. The voltage[10] actually measured is the sum of these two parts $V(t) = S(t) + N(t)$. To calculate the actual noise reduction, we need to compute the standard deviation of the average of $V(t)$ after N records have been averaged. The signal is assumed to repeat exactly from trial to trial, so $\sigma_s = 0$ while the noise standard deviation is reduced as the square root of the number of samples. With these assumptions, $V(t) = V_s(t)$ and $\sigma_v = \text{sqrt}(\sigma_n/N)$. Note how random signals can be reduced by summing signals and taking the average of signals, as shown in figure 6.14.

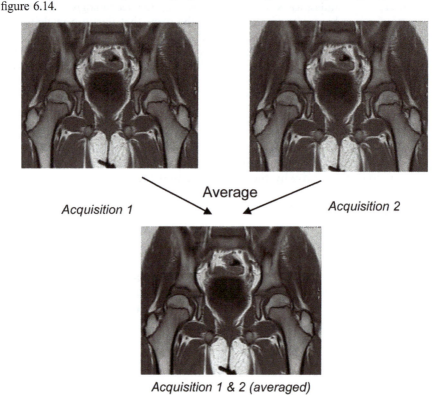

Figure 6.13. Illustration of how multiple images result in an increase in SNR of an image when the images are combined (via averaging).

---

[10] We describe the voltage $V(t)$ here as how we measure signal in MRI through the hardware. In chapter 2, we described measuring the signal through electronics. Recall the core concept from chapter 1 of the changing the magnetic field creating the electric field (i.e., current). This is an electrical signal $S(t)$ that comes from the measured voltage in the receive coil.

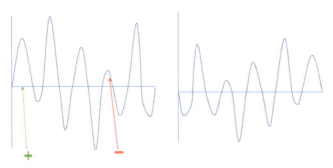

Figure 6.14. Number of Averages (N_{AVG}) refers to each time an entire pulse sequence is repeated. These graphs show two random signals, where there are two signals with some random variations that end up subtracting from each other and some that end up additive. The result, though, is an overall benefit of the square root of N.

Let's refer back to our pulse sequence diagram, as shown in figure 6.6. There are multiple phase-encoding (PE) steps within a pulse sequence diagram, as shown in figure 6.6. As a tie-in with the amount of time per scan that we discussed in section 6.2, each PE step requires the repetition time (TR) to repeat, acquiring more data each time, but also causing the duration of the scan to increase. In addition, we can repeat the entire pulse sequence, referred to in our SNR equation as the number of averages. We must account for both repeats of the sequence due to PE and averaging, hence ($N_{AVG}*N_{PE}$), as illustrated in figure 6.6, which was shown in prior section 6.2 when we discussed time tradeoffs. In this case, however, the repetition of the blocks is a form of 'averaging.'

Whitney (MRI technologist) said: "If you decrease the PE from 256 to 192, it saves some time, but the FOV might also be increased to compensate for the change in SNR. For example, on an abdomen I might increase the FOV from 40 cm to 44 cm[11]."

Another balancing act in the day of an MRI technologist!

---

[11] There was a little bit of loss of resolution related to the increase in FOV (approximately ~83% of the original signal, which is related to the ratio between 40 and 44 cm squared because it is the area in a plane (i.e., (40 cm/44 cm)^2).

6.3.4 SNR changes based on time-related effects (sampling time/receive bandwidth) and relaxation effects

Which parameters alter SNR with time? There are two main ideas here: (1) sampling time/receive bandwidth and (2) relaxation effects.

$$SNR \propto \frac{FOV_{RD}}{N_{RD}} \times \frac{FOV_{PE}}{N_{PE}} \times \Delta z \times \sqrt{N_{AVG} \times \frac{N_{PE}}{BW_{RECEIVE}}} \quad (6.5)$$

Increasing your sampling time which is (1/BW)
Increases the SNR by sqrt(BW)

6.3.4.1 Sampling time impact on SNR

Intuitively, the longer you sample, the more signal you will acquire. More signal results in a higher SNR. Recall that receiver-bandwidth has an indirect relationship with sampling time. In order to increase our sampling time, BW is decreased, as illustrated in figures 6.15 and 6.16.

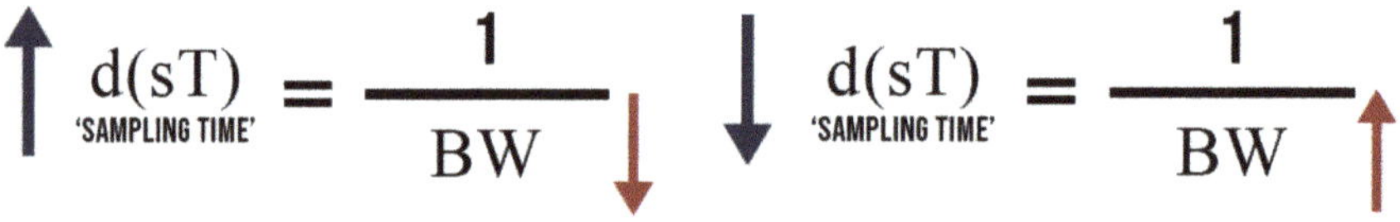

$$\uparrow d(sT)_{\text{'SAMPLING TIME'}} = \frac{1}{BW} \downarrow \qquad \downarrow d(sT)_{\text{'SAMPLING TIME'}} = \frac{1}{BW} \uparrow$$

Figure 6.15. BW is measured in the frequency realm, but it may be easier to grasp BW by thinking in terms of time. Remember that BW is mathematically inverse to sampling time.

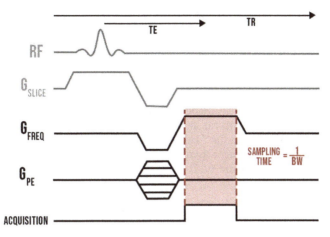

Figure 6.16. Pulse sequence diagram showing the sampling time. Note that the total sampling time is inversely related to the receiver bandwidth of the sequence (per pixel).

Natalie, can you explain receive BW and SNR in your own words?

"I understand that SNR is a ratio between the average of the signal to the standard deviation of the signal. Noise will affect this ratio, which then affects the clarity of an image. Bandwidth is the same as the inverse sampling time. To refer to the analogy used in chapter 5, it is like a window. Bandwidth and SNR are inversely proportional, so when you have a high bandwidth, you will have a smaller sampling time, which will yield a smaller SNR, which usually warrants an image with less clarity. So, as I understand, a good rule of thumb is to decrease receiver bandwidth when you can to provide appropriate SNR to a have a good scan."

Dr. Wu added: "Those are great thoughts; you have come a long way in your understanding of Bandwidth and SNR. Some implementation consequences include that smaller bandwidths sometimes lead to other challenges, including:

(a) More chemical shift misregistration with lower BW;
(b) Potentially some susceptibility artifacts increase with lower BW;
(c) Timing of sequences could be modified by changing the BW (as it could affect TR since sampling time has been modified)."

6.3.4.2 Relaxation effects on SNR

Recall that T2/T2*· is the measurement of time it takes for the 'spins' to decay after being excited, back to their natural state. The shorter the T2*/T2 time, the less signal is acquired. The opposite is true for T1. If the 'spins' recover quicker, resulting in a short T1, the signal emitted is stronger than that with a longer T1. A shorter T1 results in a higher SNR, as shown in figure 6.17.

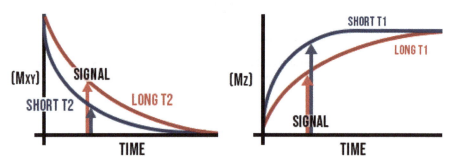

Figure 6.17. Focusing on the uprising arrows, we can compare the signal between various parameters. Short T2 = less signal. Short T1 = more signal.

A longer repetition time (TR) allows the 'spins' more time to recover, especially for T1. This typically means more signal. A shorter echo time (TE) means that you will generally have less decay in terms of signal. 'Driving'[12] the TR and TE will have an impact on the SNR (not shown in formula, for simplicity). As guidance, consider using a shortened TE and a longer TR. When selecting these parameters, be aware of side effects, such as distorting the weighting of the image or increasing the duration of the sequence.

Example: Time and pulse sequence parameters will have an impact on the relaxation effect, particularly the TE and TR. For example, a short TE will generate an image that has more signal in it, as the effect is a decay over time:

$S_1 \propto e^{-TE1/T2}$ is larger than $S_2 \propto e^{-TE2/T2}$ when TE1 < TE2

Similarly, TR represents regrowth, in which a short TR would mean a larger signal, as shown in figure 6.18:

$S_1 \propto (1 - e^{-TR1/T1})$ is shorter than $S_2 \propto (1 - e^{-TR2/T2})$ when TR1 < TR2

---

[12] Much like an automobile, manipulating parameters in MRI (such as TR and TE) are like '**driving**,' and we sometimes refer to changing the parameters in that fashion.

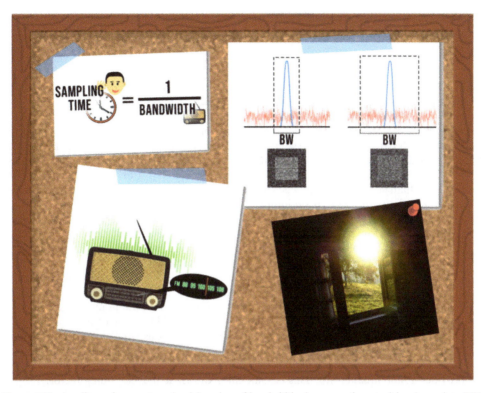

Figure 6.18. A collage of concepts and quick review of bandwidth. As we continue to delve deeper into BW, review the above images and summarize the main points from chapter 5. Recall our analogy of the 'window.' As we open the window wider, we let in more noise (air, wind). However, the signal (sunlight) remains tight and unchanged. This 'File:The open window (6028681236).jpg' image has been obtained by the author from the Wikimedia website where it was made available marcsiegert under a CC BY 2.0 licence. It is included within this article on that basis. It is attributed to marcsiegert.

Whitney (MR technologist) reflected on the role that relaxation effects (T1 and/or T2) play in the overall SNR: "It depends on which weighting your image has…it is opposite for T2 or T1 imaging."

Whitney says, "it depends on the weighting of the scan. For example, T2-weighted scans TE, which can alter the SNR. Typically, shorter TEs will increase the SNR of scans.

On a T1, if you end up using a very small TR, the signal will not recover as much, which can decrease SNR and so should be noted or compensated for."

Three other concepts of importance to SNR optimization: (a) **Flip angle (FA)**, (b) **Cross talk (overlapping effects)**, and (c) **Coil Filling Factors**

Flip angle (FA): In general, increasing the FA up to 90 degrees in a single FID or spin echo increases the overall signal. However, this is a little strange for gradient echo sequences (because they are repeated), and accumulated signal happens over multiple TR, which gives different results than expected (for incoherent and coherent signal pathways, you can refer to Brown *et al* for more details [2]). For these sequences, an angle less than 90 degrees usually generates a more maximal signal and takes a little bit of math to figure out what is optimum.

Cross talk (overlapping effects): Overlapping slices lead to signal cancellation, typically caused by imperfect slice selection. Gapping and/or interleaving of slices minimizes the effect of slice overlap. More on this will be presented later in the section on cross-talk and multislice in the artifacts section (table 6.1).

Coil Filling Factor: The impact of the coil can affect the SNR. The size and position of coil matters, since the observation is restricted to a smaller and more focused region. Typically, a smaller coil will yield more SNR, since it is not pulling in as much noise that arises from regions outside of the region of interest. This is one of the reasons for phased array coils, since you have a collection of numerous smaller coils in phased array systems that can be used to cover the region. Basically, focusing coils help increase the overall SNR, which means a coil that is the optimized size coverage and provides a more quality 'fit.' This is sometimes known as the 'coil filling factor.'

6.3.5 Knowledge check

You have a pulse sequence for which you want to double the slice thickness, double the averages, halve the pixel size (in two directions), and double the bandwidth. What is the change in signal?

$$\text{SNR} - \text{FOV}_{RD}/N_{RD} \times \text{FOV}_{PE}/N_{PE} \times -z \times \text{sqrt}(N_{AVG} \times N_{PE}/\text{BW}_{receive})$$

Answer:

Double the thickness is double the abundance (2×), double the averages is sqrt(2) times the signal. Half the pixel size, because it's in two directions, means 1/4 of the signal. Double the bandwidth means a decrease in SNR by 1/sqrt(2). The result is half of the total SNR = 2 *·sqrt(2) *·1/4 *· 1/sqrt(2).[13]

---

[13] This may seem tricky, as there is a lot to recall when doing these problems. Perhaps remembering the formula here is a structure that can aid your retention of this information on SNR and time.

Table 6.1. By increasing parameters below++Conceptual domain and/or notes

Matrix size		+	+	Spatial
Bandwidth		nc	nc	Time
Magnet strength	+	nc	nc	Spatial/abundance (parallel/antiparallel difference)
# of PE steps	+	nc	+	Averages
Oversampling (freq)	nc	nc	nc	Advanced topic[1] [4]
Oversampling (phase)	nc	nc	doubled	Advanced topic [5]

[1] These topics are presented well at MRIquestions.com, which is a great resource for understanding physics, particularly from a physician's point of view [4].

Figure 6.19. The control of body temperature is critical in many conditions. The environment of MRI is important for patient safety. Reproduced with permission from noorhaswan/stock.adobe.com.

It is important to consider the impacts of time and SNR simultaneously. What happens when you increase the number of phase-encoding steps in a spin sequence from 192 to 256 steps?

Answer:

If you increase the sequence to change your number of phase encodes (i.e., $N_{PE} = 192$) to a number of phase encodes of $N_{PE} = 256$, the time of the sequence increases by 256/192, or four-thirds = ~1.33, which also results in an increase in SNR by ~sqrt(4/3) = 1.15 times[14].

Congratulations! We have provided the collage in figure 6.18 to celebrate your accomplishment. You have made it through that fairly long section on SNR regarding spatial size and abundance, averaging, sampling time, coil positions, and cross-talk interference. Hopefully, this section will aid you in your understanding of one of the other challenges.

6.4 Tradeoffs with SAR (energy)

Maintaining body temperature is critically important for patient safety and understanding the limits is important to ensuring ensure safe and effective care. Heating in the body can be associated with patient discomfort, health deterioration, or potential burns. RF-induced temperature changes, as symbolized in figure 6.19, are of potential concern in MRI and must be managed by the MRI technologist and radiologists.

---

[14] Review table 6.1 for general guidance on the direction of increase or decrease in SNR or time in accordance with various parameter changes.

6.4.1 Fun-fact bubble

As a break from all the heavy physics, let's look at some interesting thermal images for fun.

In addition to MRI, the author is interested in veterinary projects and has collaborated to help diagnose various animals, including horses and osteoarthritic giraffes. The following images were taken with an infrared thermal camera. The signal variation differs between the hotter areas and the colder areas. Thermal differences can be indicative of blood flow patterns. Still, there is quite a way to go with this technology for the utility of diagnoses. While this does not have much to do with MRI, it does provide the context of heat deposition in an image (figures 6.20 and 6.21).

Now back to MRI! In MRI, you are imparting energy (i.e., with the RF transmitter to the subject). The application of an RF pulse imparts energy to the subject.

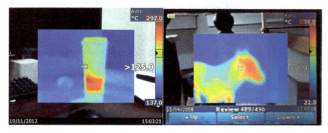

Figure 6.20. (Left) picture of a coffee cup with hot liquid in it. (Right) picture of Chico (the author's dog). Note the range of temperatures of the animal. The temperature in the head is higher than the temperature in the rest of the body, which probably relates to metabolism.

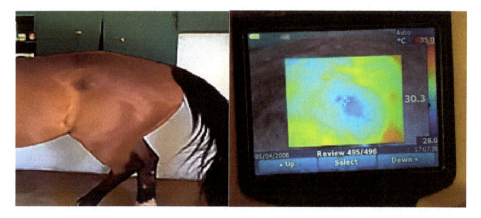

Figure 6.21. The thermal image was taken of this horse to assess blood flow due to a kick injury.

Figure 6.22. A video provided by the International Society of Magnetic Resonance in Medicine (ISMRM).

Specific absorption ratio (SAR) measures the amount of RF energy (heat) absorbed by the patient's tissue during MRI. The typical concern is to maintain values that are within the regulatory limits set by the FDA. SAR is dependent on multiple factors, such as tissue conductivity, body size, magnetic field strength, RF frequency, and duty cycle of the RF pulse[15]. The science of monitoring/estimating body temperature changes is one that is constantly evolving, particularly as we are moving toward higher field magnet systems [6]. As general guidance for RF pulses, their deposition contribution to SAR follows this formula[16]:

$$\text{SAR} \propto B_0^2 \times \text{Amplitude}^2 \times (\text{Duty Cycle}) \text{ of the RF Pulse} \qquad (6.6)$$

How do I use the above formula?

Generally, manufacturers will report the estimated SAR for the given sequence you are running directly on the scanner. However, it is useful to note that SAR goes up as the square with field strength of the magnet.

A second metric for heating factors generated by the RF pulses is B1+RMS. B1+RMS is a measure of a time-weighted average RF magnetic field exposure, calculated over 10 second intervals, measured in microtesla (μT). The value depends on flip angle, pulse type, number of echoes, slices, and TR, but does not reflect the effects on the patient. B1+RMS is the suggested metric when handling a patient with conductive implants or devices, as shown in figure 6.22. For further breakdown of B1+RMS, visit MRI-Safety.com [7].

Note, this is not quite the same as the conventional SAR measurement that was previously mentioned. The B1+RMS is a measure of the output, and not necessarily how the tissue responds. For safety purposes, B1+RMS and SAR specifications are sometimes provided by the instrument manufacturer to help keep the potential for heating under more control. It is instructive for any user of MRI to be aware of the FDA SAR limits, as well as the operating conditions.

---

[15] The duty cycle of the RF refers to a duration of exposure of the RF across the time of delivery in the pulse sequence.

[16] Note that the amplitude of the RF represents the 'height' of the RF (loosely, the mean height where the energy is deposited), while the duty cycle is a factor reflecting the width.

FDA SAR limits:
- Whole body: 4 W kg^{-1}/15 min exposure averaged;
- Head: 3 W kg^{-1}/10 min exposure averaged;
- Head or torso: 8 W kg^{-1}/5 min exposure per gram of tissue;
- Extremities: 12 W kg^{-1}/5 min exposure per gram of tissue.

  IEC (International Electrotechnical Commission) SAR limits of some European countries:
- Operating conditions are provided by agencies and modes that are used by technologists to enable them to maintain better control over their acquisition in terms of SAR.
- Level 0 (normal operating mode): whole body 2 W kg^{-1}; head 3.2 W kg^{-1}; head or torso (local) 10 W kg^{-1}; extremities (local) 20 W kg^{-1}.
- Level I (first level controlled operating mode): Whole body 4 W kg^{-1}; Head 3.2 W kg^{-1}; Head or Torso (local) 10 W kg^{-1}; Extremities (local) 20 W kg.
- Level II (second level controlled operating mode): all values are over Level I values.

Strategies to minimize heat deposition due to the MRI [8]:
- Increase the T; however, this may affect the overall scan time.
- Reduce flip angles (for FSE sequences, for example, use shorter refocusing pulses rather than 180° refocusing pulses). Note that this can alter image contrast-to-noise and/or signal-to-noise ratios.
- Reduce the number of slices in an acquisition. This may, however, lead to longer scanning times by requiring larger TRs, for example.
- Reduce the number of echoes in multi-echo sequences (for example, lower echo train length in fast spin echo-based sequences). However, this can lead to longer scanning times.
- Control the scanning room temperature and humidity (follow manufacturer specifications). Dress the patient in light clothing when possible. Note that this may affect the patient's modesty.
- Consider taking breaks between high SAR acquisitions or interleave high SAR and low SAR acquisitions to allow greater patient cooling.
- Be sure the patient ventilation system, such as the fan system in the scanner, is turned on.

  From a clinical perspective:

  Considerations for increases in body temperature that should be made for patients include the following [9]:

  - cardiovascular disease;
  - hypertension;
  - diabetes;
  - increased age;
  - obesity;
  - fever;
  - pregnancy (risk for fetal heating);

- drug regimens that may affect thermoregulatory capabilities (e.g., diuretics, tranquilizers, vasodilators).

Finally, let us discuss general guidance in accordance with the above tables. The consideration of SAR is becoming more important as more scanners are moving to higher fields (note the formula above in regard to the potential increase in SAR at higher fields, including 3T, 7T, and potentially higher fields soon). There is also a trend to consider conditions other than the pure calorimetric model and to use a B1 +RMS. The author continuously compiles a current list of resources in the area as a method to try to keep up with this advancing and growing field of concern. Refer to the American College of Radiology's (ACR) release on MR Safe Practices from 2019 for updated safety concerns [10] and continue to watch for future updates by the ACR and other MR safety interested organizations.

Carlie (premed) reflects on SAR (specific absorption ratio): "I understand SAR as the amount of heat being deposited into the patient. The ratio depends on the size of the patient, magnitude of the scan, and other factors. I also learned that it is important to monitor this carefully, especially for patients with implants or comorbid conditions. If the patient is warming up, it could be because the scan is depositing too much heat into their body tissue."

Whitney (MR technologist) says: "Off the top of my list is to be concerned most about heating in older people. However, we manage the heating in everybody. Monitor to see if they have elevated body temperature. Also remember with blankets on them or clothing that may restrict body cooling. Consider extremity vs. core body temperatures for management. Also, remove metal when possible to decrease risk of burns.

Always keep heating present in your mind, as it is a safety issue. It is also an incredibly important issue when it comes to MRI safety when there are implants. There are plenty of examples (at many sites across the country) where patients have been burned by the scanner, and these need to be controlled."

6.5 Knowledge check

(1) General guidance for RF pulses and their deposition with the following formula:

$$SAR \sim B_0^2 \times \text{Amplitude}^2 \times \text{Duty Cycle}.$$

(2) Highlighted FDA SAR limits, as well as the operating conditions.
(3) Tips provided in clinical strategies to minimize the heat deposition due to the MRI that can include reducing the flip angle when possible, altering the transmit bandwidth of the RF pulse, and/or elongating the TR.

(4) Physiological parameter adjustments for increases in body temperature, including reducing the scanning room temperature, increasing the rate of the fan that blows on the patient, and dressing the patient in light clothing, should be made.

6.6 Summary of MRI tradeoffs

We restate the 'you can do things fast, cheap, or good' saying here as a metaphor for tradeoffs in MRI.

- Fast (scan time)—Remember that optimizing speed will improve your clinical efficiency. It is essential to reduce the time of scan to help reduce effects from patient movement. To do so, it is important to consider factors such as averaging, number of phase encodes, and the repetition time, as well as the particular types of sequences you are using (see figure 6.23).
- Cheap (SAR)—SAR is a metric that the machine uses. The cost here is a safety factor, especially at higher field scanners, and has contributions from a myriad of sources, including the length and strength of the RF pulse.
- Good (SNR)—As we learned in the chapter, the size of the voxels matter in SNR, and the number of samples and amount of sampling time can also affect the SNR. Qualitatively, it is helpful to think of the effect of increase in SNR as improving the overall 'confidence' in the image.

All of these are important ideas. To date, not all manufacturers are able to provide scanners with automated visualization of parameter choices in their scan interfaces. However, guidance by the scanner, especially in terms of manipulating or optimizing parameters, has potential to have a greater and greater role over the increasing number of controls and the interactions of different parts of MRI machines. The author is an inventor of a patent on an operating calculator that encompasses aspects of SAR, relaxation, SNR, and time [11], and has worked with technologists and physicians to optimize the selection of these parameters, as observed in figure 6.24.

Figure 6.23. Make tradeoffs with intention. A good motto for MRI.

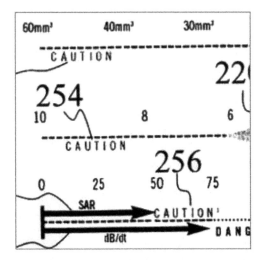

Figure 6.24. Optimizing parameters, including SAR, relaxation (CNR), time, and SNR, to create factors as built into an interface. The goal is to create a desirability and/or monitoring function that after sliding parameters illustrates modifications in factors that permit the users to better understand the tradeoffs made by changing parameters when setting up scans. Reproduced with permission from Wu and Havens [12].

It can be advantageous to have these factors on our machine through a relative SNR calculator, a total time, and the SAR (the latter will either trigger a warning if limits are close to being exceeded and/or potentially a full stop of the system if operating conditions are exceeded), especially in an age with the assistance of technologies such as artificial intelligence and machine learning (ML).

Regardless, congratulations! Arriving at this point in the book, you have already mastered a lot by improving your understanding that includes balancing SNR, time, and energy (heating) tradeoffs. This will provide the reader with an advantage when it comes to manipulating the parameters on their own scanners for developing MRI protocols, as well as for combating artifacts. The goal is to create the opportunity to improve 'well-being' for the patient through an efficient organization that also minimizes the likelihood of breaching conditions of safety and risk.

References

[1] Bernstein M A, King M F and Zhou X J 2004 *Handbook of MRI Pulse Sequences* (Oxford: Elsevier) pp 267–97
[2] Brown R W, Cheng Y-C N, Haacke E M, Thompson M R and Venkatesan R 2014 *Magnetic Resonance Imaging: Physical Principles and Sequence Design* 2nd edn (New York: Wiley)
[3] Koch K M *et al* 2021 Analysis and evaluation of a deep learning reconstruction approach with denoising for orthopedic MRI *Radiol.: Artif. Intell.* **3** e200278
[4] Frequency Wrap-Around (n.d.) *Questions and Answers in MRI* (https://mriquestions.com/frequency-wrap-around.html)

[5] Phase oversampling? (n.d.) *Questions and Answers in MRI* (http://mriquestions.com/phase-oversampling.html)
[6] FDA 2021 MRI information for industry (https://fda.gov/radiation-emitting-products/mri-magnetic-resonance-imaging/mri-information-industry)
[7] Safety topic/article (n.d.) (http://mrisafety.com/SafetyInformation_view.php?editid1 = 362)
[8] Allison J and Yanasak N 2015 What MRI sequences produce the highest specific absorption rate (SAR), and is there something we should be doing to reduce the sar during standard examinations? *Am. J. Roentgenol.* **205** W140–140
[9] Golestanirad L *et al* 2019 Changes in the specific absorption rate (SAR) of radiofrequency energy in patients with retained cardiac leads during MRI at 1.5 T and 3 T *Magn. Reson. Med.* **81** 653–69
[10] Greenberg T D *et al* (ACR Committee on MR Safety) 2020 ACR guidance document on MR safe practices: updates and critical information 2019 *J. Magn. Reson. Imaging* **51** 331–38
[11] Wu D H and Havens T K 2003 System and method of user guidance in magnetic resonance imaging including operating curve feedback and multi-dimensional parameter optimization *World Intellectual Property Organization Patent No. WO2003021284A1*
[12] Wu D H and Havens T K 2003 System and method of user guidance in magnetic resonance imaging including operating curve feedback and multi-dimensional parameter optimization *Patent* W02003031284A1

IOP Publishing

MRI: Connecting the Dots
A start to concepts
Dee Wu

Chapter 7

MRI Artifacts (clinically relevant)

(Geometric distortion, partial voluming, slice cross-talk, dielectric effect, aliasing, truncation artifacts, zipper artifacts, moiré artifacts, ghosting, chemical shift artifacts, magic angles)

7.1 Introduction

Diagnosing the presence or absence of disease in patients has been described as a process that identifies a set of categories agreed upon by medical professionals [1]. The discussion of the cause of artifacts is often one of the main requests by attendings, residents, and technologists, particularly in regards to board preparation. This chapter describes MRI artifacts, as it is important to radiological imaging. MRI artifacts refer to something seen on an image that is not present in reality, but appears due to idiosyncrasies of the MRI hardware and/or software. More often, artifacts describe findings that are due to phenomena occurring outside of the patient's own anatomy that may obscure or distort the image.

Occasionally, artifact accentuation is intentional because the artifacts may be advantageous to the interpreter, making anatomy/pathology easier to appreciate (as we described for susceptibility effects, which can be used for detecting hemosiderin and/or calcifications). As an example, for arteriovenous malformations (AVM) that have bled, there can be residual iron, which would be a sign of hemosiderin staining. The altered signal due to blood can be visualized using an iron-sensitive sequence, such as gradient echo T2*-weighted sequence.

As an interpreter of imaging, it is important to be aware of the main artifacts of the examination being reviewed to avoid issuing an erroneous report [2]. In radiology, it is important to provide diagnoses accurately and in a timely manner, particularly as it can aid the team with a greater opportunity to achieve better health outcomes. Of concern is that artifacts can contribute to misinformation. Being able to understand these artifacts will aid in ensuring patient safety (from mistreatment) and well-being (establishing health status in a timely fashion). Thus, academic radiology departments

Figure 7.1. Danger! Artifacts can be the source of false positive or false negative diagnoses. Interpret those images with caution.

are tasked with training residents to recognize MRI artifacts (a topic sometimes emphasized highly on the boards, and so we have created this 'caution' sign, as shown in figure 7.1) to minimize the risk of potential complications.

We will divide this chapter into three parts. In part one, we cover a group of **general artifact concepts**, starting with geometric distortion and ending with the concept of aliasing. In part two, we proceed to **'frequency' domain artifacts**, which encompass truncation artifacts to ghosting. Finally, in part three, we discuss the impact of **material artifacts** (chemical shift of fat and effects of imaging collagen in some locations) that are important to MRI. We will proceed to discuss the conceptual basis of each artifact using four sections: (1) description; (2) effects/appearance; (3) examples; and (4) remedies/improvement strategies. At the end of the chapter, we will provide a summary of all of the artifacts. While there are multitudes of artifacts to cover, we will describe the most common MRI artifacts that you may encounter.

7.2 Starting with initial concepts on artifacts

To start the discussion on artifacts, we begin with an idea that we previously discussed in chapter 3 when we compared T2* with T2. This concept involved inhomogeneity, where two dissimilar materials in close proximity cause bending of magnetic waves, thus creating potential artifacts.

7.2.1 Distortion (geometric)/susceptibility artifact

Description: Geometric distortion is the warping and/or deformation within the image due to local variations in the field, as illustrated in figure 7.2. Local gradient fields are distorted due to the change or bend in direction of the magnetic field as it passes through different interfaces.

Effects/appearance: Distortion affects both the phase-encoding and frequency-encoding directions on images. Gradient-based sequences are also more susceptible

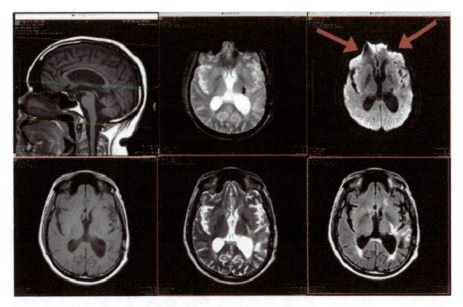

Figure 7.2. Note the variation in geometric distortion present throughout these series of images. The variation is most prominent on the echo planar scan (upper right, red arrows).

to distortion than are spin echo-based sequences. The appearance is a shift in signal position and/or signal loss.

Remedy/improvement: If using gradient echo pulse sequence, consider switching to another sequence that uses spin echoes or a propeller sequence to reduce the amount of geometric distortion in the sequence. Another option is to apply long Echo Train Length (ETL) in Fast Spin Echo (FSE) sequences for a further reduction in the effects of susceptibility artifacts. For instance, those effects would occur in the presence of metal in orthopedic imaging[1], such as with knee braces, surgical screws, or hip implants [3]. This can be used sometimes to aid with metal implants and the artifacts they can cause.

Whitney (MR technologist) reflects on how you deal with susceptibility and whether she has had to deal with that: "It really depends on what you are scanning. You can run "sat" bands in front or behind, you increase BW to help, try to switch from 3T scanner to a 1.5T scanner, and run STIRs to avoid conventional fat sat."

---

[1] A FSE sequence applies multiple 180 spin echo pulses in a single TR. The number of echoes in the FSE is known as the Echo Train Length. This concept is advanced; for curious readers, see references [3, 4].

7.2.2 Partial voluming

Description: Partial voluming occurs when there is more than one tissue type within a voxel, resulting in poor spatial resolution [3]. This is particularly emphasized by choosing slices that are too thick, as shown in figure 7.3.

Effects/appearance: The partial voluming artifact can occur in any direction (phase, frequency, slice of the pulse sequence) for any sequence. The amount of partial voluming is based on the size of the anatomical structures and the acquired resolution in that direction, as shown in figure 7.4.

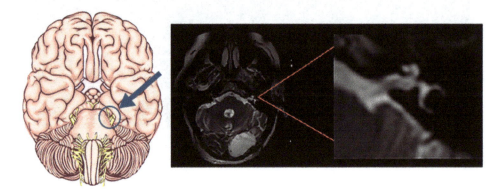

Figure 7.3. Thick slices, where it is hard to see the nerve roots on these images.

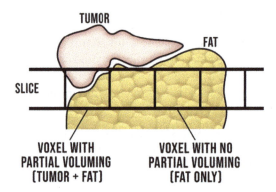

Figure 7.4. Larger voxel size increases the chance of having multiple tissues within a single voxel. Adobe Stock Images: © Zerbor.

Remedy/improvement: Implement sequence with smaller voxels through either use of a smaller field of view (FOV) image or a higher matrix, for example changing from a 256 matrix to 512 matrix sequence, which reduces the pixel size and improves resolution. Another alternative is to acquire thinner slices in your MRI acquisition.

Nick P. (third-year medical student) responding to whether he had ever thought about partial voluming: "We see this often in terms of resolution, particularly with small nerves that we are trying to observe. Having reduced partial volume can mean a lot for radiologists, as it affects the visibility of structures such as the tiny cochlear structures and nerve roots, making it much more difficult to get an accurate and detailed read."

7.2.3 Multislice cross talk

Description: Slice selection does not produce perfect shapes[2]. In reality, the slices will 'bleed' into one another if not spaced out. The overlapping causes the RF pulses of adjacent slices to cancel each other out, resulting in signal loss, as shown in figure 7.5.

Effects/appearance: Signal loss (and potential artifacts) are due to sequential overlapping slices in 2D sequences, as shown in figure 7.6. There may be potential for loss of contrast if there is crosstalk present in the acquisition.

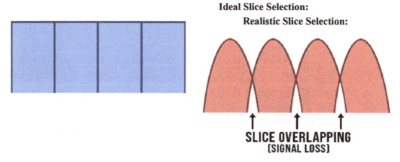

Figure 7.5. Realistic shape of slice selection.

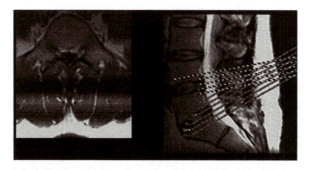

Figure 7.6. The effects of cross talk can be visualized in a myriad of ways, one of which is shown in the figure where the slices overlap on top of each other, resulting in signal loss.

---

[2] Slice selection is created by a time-limited duration of RF pulse. See chapter 5 on slice selection for images illustrating the shape of the imperfection.

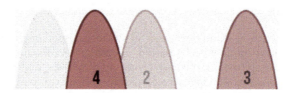

Figure 7.7. Slice interweaving builds gaps within slice selection to avoid signal loss. As the effects of acquisition from '1' and '2' fade, we are able to return and fill the gap. Animation available at https://doi.org/10.1088/978-0-7503-1284-4.

Solution/improvement [5]: Check whether slice interleaving, an option on the MRI scanner, has been turned on. Slice interleaving alternates acquisition between every other adjacent slice, then returns to acquire the intermediate slices, which is perhaps better seen in animation mode, shown in figure 7.7. This allows enough time for T1-decay to minimize signal interference between adjacent slices.

Natalie's reflection: "I think I understand what cross talk is. It is when you the slices are taken at the same time and can overlap and interfere with each other. What are some things you can do to reduce the chances of cross talk?"

Whitney (MRI Technologist) said, "The major time I see when we have to be careful of this is in spine imaging, especially through disk spaces through the spine. We have to make sure the slices aren't crossing and have a big enough slice gap.
We encounter this issue because of the proximity of slices, especially when imaging the spine. While turning on interleaved slices does help occasionally, you may run into the problem where the effect of one slice impacts another. This creates either artifacts or signal reduction, which should be avoided."

7.2.4 Dielectric effects

Description: RF energy, as shown in figure 7.8, is absorbed differently across the body, leading to an inhomogeneous B1 field. Variegated RF penetration leads to 'shading' and intensity variation across the image [6].

Effects/appearances: Dielectric effect artifacts become more prominent with increasing field strength, say 3 T versus 1.5 T. The argument that these artifacts are due to dielectric effects is based on considering RF-wavelengths in tissues as a function of field strength. See figure 7.9; as field strength intensifies, the RF wavelength becomes shorter. This becomes an issue when the anatomical regions being scanned are larger than the wavelength. For example, consider the effect of ascites (fluid buildup in the abdomen, which can be particularly relevant in the peritoneal cavity), cirrhosis, or pregnancy, which increase the size of the abdomen outside the norm. Smaller RF waves are more likely to interact, resulting in constructive and deconstructive interference. Central brightening has been demonstrated in some cases, as shown in figure 7.10 [6].

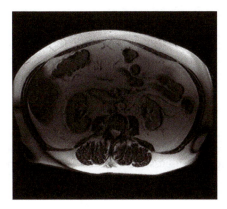

Figure 7.8. Variation of signal intensity in abdominal scan.

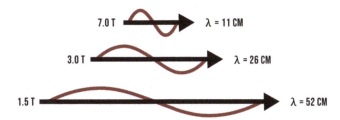

Figure 7.9. Wavelength of RF energy decreases as field strength increases.

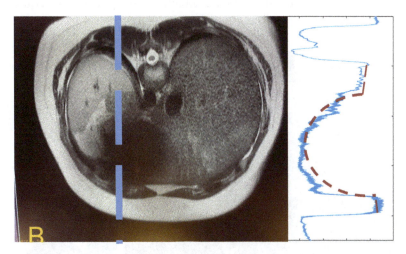

Figure 7.10. 'Dielectric' artifact at 3.0 T. Center of abdomen profile dips toward center, exhibiting a standing wave-like structure.

Remedy/improvement: Drain ascites (if present in the patient) prior to imaging, when possible. Consider using less than 3 T systems if dielectric effects are suspected to be a problem. Potentially enable the use of transmit sense (multiple transmitter technology, which can be used to assist in providing better spatial distribution of the RF transmission) [7].

Whitney (MR technologist) was asked, "Have you ever tried to use dielectric pads?" She said: 'Not at 1.5T, but maybe it could be useful at higher fields. We just make sure that the ascites is drained."

Dr. Wu also added "While I don't commonly see the use of dielectric pads, I do know that they potentially have the ability to help. This is commonly discussed by our technologists. However, dielectric effects cause standing wave issues at higher fields. This is problematic and needs to be monitored and addressed."

7.2.5 Aliasing

Description: The aliasing artifact is also referred to as wrap-around artifact. Technically in MRI, the encoding of 'frequency space' is applied in a periodic manner. When the period does not have a high enough sampling rate, the frequency is erroneously assigned. This undersampling in frequency causes the image to create a shifted copy within the FOV, as depicted in figure 7.11.

Effects/appearances: The result is an object wrapping from one side of the FOV to the other. Aliasing happens in the phase direction (as opposed to frequency). The effect can happen with any pulse sequence.

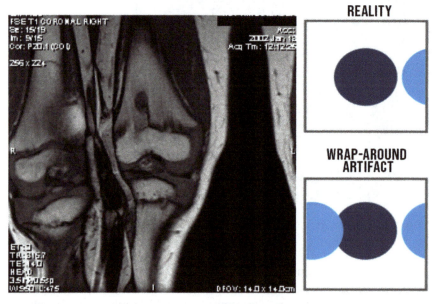

Figure 7.11. FOV is set to only scan the right patella (kneecap). Due to aliasing, the left patella out of FOV is unintentionally scanned and wraps around on top of the intended knee.

Solution/improvement:
- enlarge the field of view (FOV);
- use pre-saturation bands on areas outside the FOV;
- anti-aliasing software;
- switch the phase and frequency directions;
- switch to a surface coil, which will reduce the signal outside of the area of interest.

Whitney's prompt: Have you had to repeat scans due to aliasing? "We have to make sure your FOV is wide enough in the correct plane and turn no-phase wrap [8]. While aliasing happens infrequently, it is possible based the rate of completing scans. If the phase oversampling parameter is activated, aliasing can be prevented in the image. Additionally, selecting the right field of view can prevent aliasing. This an important part of a technologist's job, and is certainly annoying for physicians when they have to read these images."

7.3 Advanced ideas that are more concerned with frequency space

In this section, we will discuss artifacts within the frequency domain. A conceptual understanding of frequency will improve the reader's comprehension of the frequency impact in MRI.

7.3.1 Truncation/gibbs

Description: Truncation is also called Gibbs ringing. The MR image is reconstructed from k-space, which is a finite sampling of the signal subjected to inverse Fourier transform in order to obtain the final image. At high-contrast boundaries (a jump in signal or a discontinuity in mathematical terms), the Fourier transform corresponds to an infinite number of frequencies. Since MR sampling is finite, the discrepancy is manifested in the reconstructed image in the form of a series of lines, as shown in figure 7.12.

Effects/appearance: These lines can appear in both phase-encode and frequency-encode directions. While often manifested along edges, an anatomical area particularly subjective to truncation is the spine, leading the spine to appear as a potential syrinx[3]. Note, the ringing occurs parallel to structures with abrupt or intense changes, such as the skull–brain interface, as shown in figure 7.12.

Solution/improvement: Increase matrix size in the offending direction. The more encoding steps, the less intense and narrower the artifacts [9].

---
[3] Syrinx is a focal dilation of the central spinal canal, which can lie within the spinal cord itself.

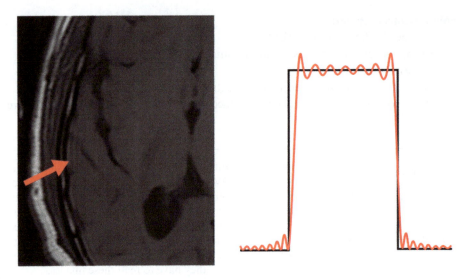

Figure 7.12. Illustration of Gibbs ringing. Animation available at https://doi.org/10.1088/978-0-7503-1284-4.

7.3.2 Zippers

Description: Appears as spurious bands that can occur in both directions. One manifestation is caused by not closing the room door properly, or equipment emitting frequencies in the room (like from wireless contrast injector equipment that may have RF-emitting elements). This is due to RF entering the scanning room from electronic equipment and being picked up by the receiver chain of imaging subsystems, as shown in figure 7.13.

Effect/appearance: The zipper artifact appears as a bright signal streaking along a line. The effect can occur in the phase direction (most common) or frequency direction (less common), which can happen through stimulated echoes.

Remedy/improvement: The zipper artifacts that are most easily controlled are those that occur when the door is open during acquisition of images. RF from some radio transmitters will cause zipper artifacts that are oriented perpendicular to the frequency axis of your image. Frequently, there is more than one artifact line on an image from this cause, corresponding to different radio frequencies. Strategies to avoid other equipment and software problems are:

1. Make sure the MR scanner room door is closed during imaging.
2. Remove all electronic devices from the patient prior to scanning.
3. If the artifact persists despite all nearby electronic equipment being turned off, check whether RF shielding of scanning room is operating as expected (no leaks or inadvertent signals generated in the room).
    a. An 'RF leak'[4] can occur between the door and the jam, and these locations may need to be cleaned or repaired.

---

[4] RF Leak results from RF waves being able to enter the scanner room, which can create artifacts in the image.

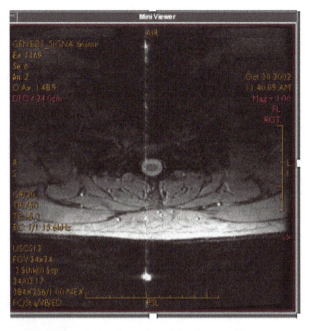

Figure 7.13. Illustration of the zipper artifact.

b. Have the vendor service team check the penetration panel, where the cables enter the room where RF leakage can also occur.

This section is reproduced from [17].

Whitney (MRI technologist) said: "If you see zipper artifact, which may be caused by 'electronic noise', then I like to look in the room to see if there is unintended electrical equipment plugged in and left turned on. The door can be left ajar and/or cracked open, which also can produce this artifact …"

Dr. Wu also added: "Zippers are notorious because sometimes they are intermittent. Occasionally it is something in the wall, and occasionally it is something in the equipment. We've had many times when the physicist, the technologist, and the service engineers have tried to track this down. It does happen and needs to be addressed."

7.3.3 Corduroy artifact (spikes)

Description: To understand this phenomenon, think of introducing a dot in k-space. This dot will produce a spectral wave across the image. These electromagnetic spikes, or 'dots,' are caused by external forces produced by malfunctioning equipment such as gradient coils. Also known as spike artifact, crisscross artifact, or herringbone artifact, as shown in figure 7.14.

Effects/appearances: Uniform stripes appear to cover the entire image, and can be vertical, horizontal, diagonal, or a combination. The artifact covers the entire image in a single slice or multiple slices.

Remedy/improvement: Call service to have them check connectors and check if it is possible that spike-detection hardware may be malfunctioning.

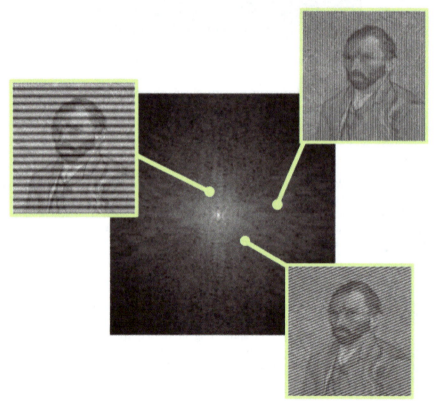

Figure 7.14. The effect the electromagnetic spike has on the image is dependent on the location of the spike within k-space. Multiple spikes may present simultaneously.

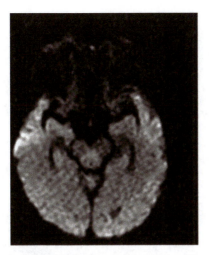

Figure 7.15. Brain scan with corduroy artifact.

Whitney (MRI technologist) reflected on when she's seen corduroy (spike) artifacts: "I've seen corduroy (spike) artifacts, and we immediately make sure that the coil connectors are tight and then, if needed, we call our MRI service providers to see if there are possible loose connections."

Dr. Wu also added: "When this artifact is seen, technologists are concerned and radiologists are annoyed by it and have to read images that they don't have complete trust in. It is important to work with service to identify whether there are sources for this artifact. Many times it could be a loose cable, particularly the gradients that have shaken loose over time due to vibration, but work carefully with your service people and, if you have an MR physicist available, have them involved, too, so they can work directly with the service engineer from the companies."

7.3.4 Moiré fringes

Description: The combination of aliasing artifacts and field inhomogeneity can lead to moiré fringes or zebra artifacts, which are interference patterns of superimposed images that add together and subtract [10], as shown in figure 7.16.

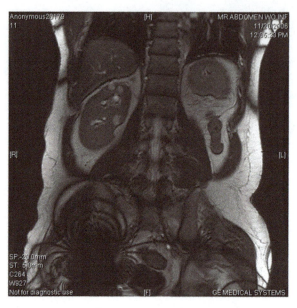

Figure 7.16. The moiré artifact is shown in this figure. Note bands in the lower left of the image, lower right, and around the armpit area of the patient.

Effect/appearance: The effect is most commonly observed when there is a combination of aliasing and field inhomogeneities effects. These are especially seen in gradient echo techniques, likely in the phase encoding direction.

Remedy/improvement: Increase FOV in the offending direction and/or use spin echo.

Whitney reflected on when she's seen moiré artifacts: "If you are looking at a hip replacement, with a large field of view, and it's going outside the coil coverage, it can happen. If so, you need to adjust the field of view or adjust the coil. Also, this artifact sometimes appears when large fields of view are needed to accommodate large habitus patients. It is pretty recognizable as an artifact. Sometimes you have to work your way through this the best you can because there are limited options. However, to recognize its sources may be helpful for the reading and for technologists to take action."

7.3.5 Ghosting

Description: Phase encoding a replication can be produced by pulsatile or alternating motion, such as vibrations. Also known as phase-encoded motion artifact, it is one of many MRI artifacts occurring as a result of tissue/fluid moving during the scan. It manifests as ghosting in the direction of phase-encoding, usually in the direction of the short axis of the image (i.e., left to right on axial or coronal brains, as shown in figure 7.17.

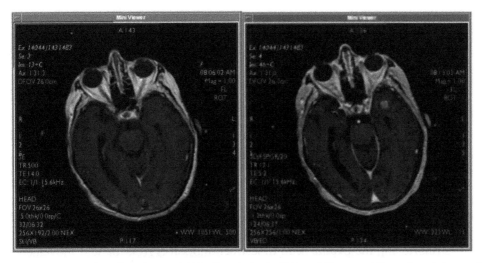

Figure 7.17. A notable artifact that overlaid itself on the brain. This could be potentially read as a lesion, such as a tumor, by an inexperienced reader. The flow artifact arises from the carotid.

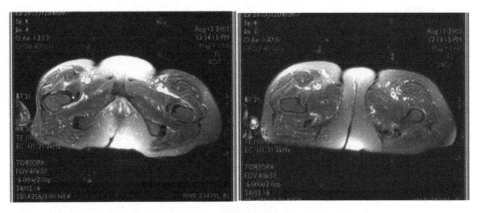

Figure 7.18. Pulse artifact as shown in the body.

Effects/appearance: These artifacts may be seen from arterial pulsations, swallowing, breathing, peristalsis, and physical movement of a patient. When projected over anatomy, it can mimic pathology, and must be recognized. Motion that is random, such as the patient moving, produces a smear in the phase direction. Periodic motion, such as respiratory or cardiac/vascular pulsation, produces discrete, well-defined ghosts. This is also illustrated in figure 7.18 in the anterior to posterior direction of the axial abdomen. The spacing between these ghosts is related to the repetition time (TR) and the frequency of the motion.

Ways of identifying phase artifacts include:
 1. Identifying known moving/flowing structures and noting that the artifact is in line with them (horizontal or vertical, depending on phase-encoding orientation).

2. Matching the shape of the ghost to that of a flowing vessel (e.g., round pseudolesion due to aorta ghost).
3. Wide windowing to see a repetitive ghost beyond the confines of anatomy.

Remedy/improvements: Patient compliance, saturation bands, reduce likelihood of mechanical or patient vibratory motion when/if possible.
1. Apply cardiac/respiratory gating;
2. Spatial presaturation bands placed over moving tissues (e.g., over the anterior neck in sagittal cervical spines);
3. Spatial presaturation bands placed outside the FOV, especially before the entry or after the exit slice, for reducing ghosting from vascular flow: arterial and venous;
4. Scanning prone to reduce abdominal excursion;
5. Switching phase and frequency directions;
6. Increasing the number of signal averages;
7. Shortening the overall scan time when motion is from patient movement.

This section is reproduced from [18].

Whitney (MRI technologist) said: "Sometimes if the artifact is going through the region of interest, I will try to swap frequency and phase so we can get right of it, especially in cases of involuntary motion like intestines, flow and swap to see if that is the cause."

Dr. Wu also added: "This artifact is one of the most difficult and potentially concerning ones. Flow artifacts can often appear to be like other things, such as masses, i.e., tumors, and it is particularly tricky when you have metastasis in the brain, which also appears small and like little dots. It's easy to interpret these little dots as being artifacts from flow and to interpret these artifacts as potentially being masses. Therefore, the understanding of pulsation artifacts is a key and critical understanding in MRI for radiologists. If you are a technologist and want to help, there are many things that you can possibly do if you are screening these images as they are coming out. If you see these flow artifacts and you want to help your radiologist, you may recommend doing an additional scan that flops the phase and frequency directions. This will help your radiologist. Of course, these extra scans add time, so work out a policy or procedure and work closely with your radiologists to maximize the productivity of your center."

7.4 Material control

7.4.1 Chemical shift (Type 1)

Description: As electrons orbit the nucleus within a molecule, a small magnetic field is generated that opposes the main field (B_0). Depending on the molecule's properties, the electrons will resonate at various frequencies [10]. Consider fat versus water. Molecules of fat are composed of long-chain triglycerides, generally having a uniform electron cloud [11]. However, water molecules display an imbalance of electrons, causing the electron cloud to only shield the oxygen molecule. This imbalance of shielding leads to varied resonance frequencies, ultimately causing the molecules to interact differently with the magnetic field [7].

Recall that spatial position is dependent upon **frequency encoding**. Note that this chemical shift is due to the differences between resonance frequencies of fat and water, in which the fat we observe resonates at a slightly lower frequency than water. If we determine spatial position off the resonance frequency of the water, then the fat becomes mismapped due to the differences in resonance frequency between the different substances. This will potentially cause signal overlaps in one direction (bright signal) and signal voids (dark signal) in the other. The difference in frequencies is extremely small, on the order of ~3.4 parts per million (ppm), which you can estimate knowing the field strength and the frequency bandwidth of the acquiring sequence. The effect is observable in both gradient echo and spin echo sequences [12], as shown in figure 7.19.

Effect/appearance: Artifactual white or dark bands, of a few pixels in width, on either side of an anatomic object (bright on one side and dark on the other).

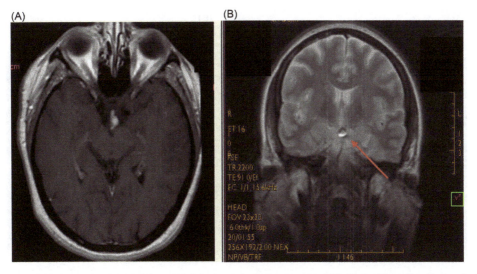

Figure 7.19. (A) On T1, a dermoid cyst is bright due to its fat components [13]. (B) On T2, notice the chemical shift artifact (see red arrow) [14]. Note that the frequency encoding direction is denoted by the green box to be in the cranial and caudal direction.

The artifact is most seen around water-containing structures. We think of it primarily on spin echo, but it also can affect gradient echo images. This is opposed to chemical shift of the 'second kind,' which is described in the next section, and which only appears on gradient echo images [15].

Remedy/improvement: The amount of shift is due to the frequency bandwidth used. Increase the receive bandwidth per pixel to reduce this artifact. However, radiologists are able to incorporate the understanding of chemical shift into their interpretations.

To estimate the chemical shift, take into consideration the receiver BW and frequency-encode matrix. For example, if the FOV BW is 32 kHz with a matrix of 256, the BW per pixel is 32,000/256 = 125 Hz/pixel. The fat-water frequency difference at 1.5 T is 220 Hz. The amount of chemical shift will be (220 Hz)/(125 Hz/pixel) = 1.8 pixels. Notice that main magnetic field strength impacts the degree of chemical shift.

7.4.2 Chemical shift (Type 2)

Description: Also known as the India ink or black line artifact. While Type 1 concerns frequency-encoding, Type 2 concerns phase-cancellation. This form of chemical shift occurs exclusively in gradient echo sequences (GRE), while Type 1 is mainly observed in spin-echo sequences. During GRE, the fat and water protons go in and out of phase with one another. Voxels containing fat and water that are **out-of-phase** appear dark due to deconstructive interference, ultimately resulting in a black outline surrounding organs, while in-phase gives an additive signal between water and fat, as shown in figure 7.20 [15].

Recall from chapter 1 how phase and frequency are related. Phase concerns the offset between waves, while frequency is the rate. If frequency between two signals varies, the phases are constantly altering between 'in-phase' and 'out-of-phase.'

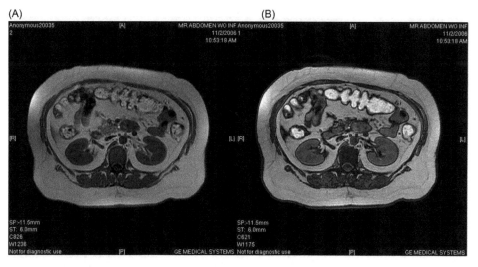

Figure 7.20. Dual Echo Sequence (A) in-phase, provides additive fat and water. (B) out-of-phase, (note the black line around the image) provides subtraction of fat and water.

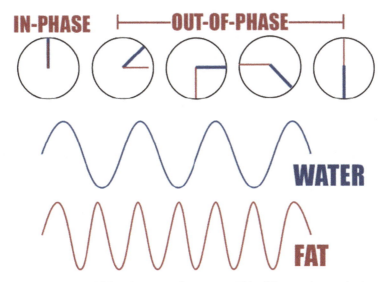

Figure 7.21. Fat rotates at a higher frequency than water. This difference in rate leads to the phase-cancellation. Depending on where you 'listen' (acquire your signal), the signal waves will either be additive (in-phase) or deconstructive (out-of-phase).

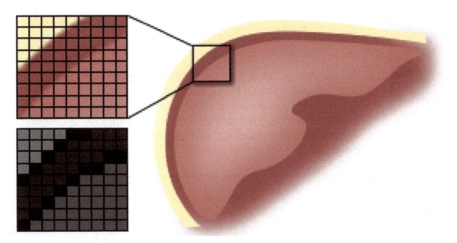

Figure 7.22. Consider an abdominal scan of the liver. Voxels containing both fat and water will be subject to phase-cancellation, resulting in a black outline along boundaries.

'In phase' is represented by water (W) and fat (F) adding at a specific time of acquisition (F + W), while 'out of phase' is described by two amplitudes occurring at different times (F − W), as shown in figure 7.21. In-phase and out-of-phase imaging can be useful in certain applications, as shown in figures 7.20–7.22.

Effect/appearance: Disease may occur near the boundaries between tissue types. It is necessary to have both looks (in and out of phase) to both delineate the edge and assess the boundary of the organs. Because chemical shift type 2 artifacts are tied to phase-cancellation and are not limited to frequency-encoding, they can occur in any direction along a fat–water interface. Dual-echo GRE sequence is commonly employed in abdominal MRI, as portrayed in figure 7.22.

Remedy/improvement: In-phase and out-of-phase are used as two different 'looks' in MRI to help with evaluating disease. Be aware that one is subtractive and the other is additive. Typically, at 1.5 T, ~2.25 ms TE is out-of-phase, while ~4.5 ms TE is in-phase. Note the TE times are different at 3 T, meriting a more complex discussion about which echo to take first (in- versus out-of-phase). It may be useful to have a discussion with your attendings/technologist about this issue. In general, you can follow a similar line of thinking at higher fields when using dual-echo sequences.

Whitney (MR technologist) said: "I've sometimes tried to increase BW to reduce the artifacts. Also, we make notes and tell the radiologist. In the abdomen it happens a lot and is normal, if they have too much iron, or they have fatty liver. Sometimes unusual-looking images are because the patient has some abnormality (just how the patient is), and it doesn't have to do with our quality. But, in these cases, it may be good to consult with your radiologist or a senior technologist."

Dr. Wu adds: "Chemical shift artifacts can be both helpful and a detriment. We definitely use this artifact and dual echo imaging, in-phase imaging, and out-of-phase imaging to help delineate organs and parenchyma. To better identify where there's fat, it is important to recognize when chemical shift happens, as the location of fat can be shifted."

7.4.3 Magic angle

Description: The reason for this change can be attributed to quantum mechanics and simply described by equations for dipole interactions within a main magnetic field. This may sound initially complex. However, it is a set of equations that describe the interaction of spins that are orientation-dependent in MRI. Normally, these orientations are averaged as protons thermally tumble around. In the case of structured collagen, lots of water binds to the outside of the collagen protein, and therefore exhibits an orientation-dependent effect [16], as shown in figure 7.23.

Appearance: This artifact occurs on sequences with a short TE (less than ~32 ms, particularly in T1W sequences and PD sequences). It is typically found in regions of tightly bound collagen at 54.74° from the main magnetic field (B_0). The effect appears hyper intense, and thus has the potential to be mistaken for tendinopathy.

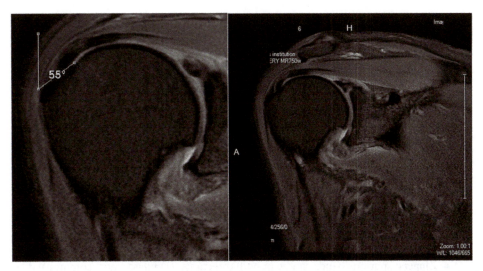

Figure 7.23. Example of magic angle artifact in a shoulder. Note that as the incident angle of the structure is approximately 55 degrees, the artifact (altered signal) could be misinterpreted as pathology, such as tendinitis.

Remedy/improvement: Compare the T1-weighted sequences, which may exhibit the effect, to those of the potentially less affected T2-images. Observe whether the result is at the ~55 degree angle with respect to the field, which is the direction of most likely enhancement in the image when the magic angle artifact is observed [16].

 Dr. Wu reflected on a recent ankle tilting protocol on which he is working with one of the MSK radiologists: "Magic angle is something that some MSK radiologists begin to learn about, know about, and try to read through, occasionally when we are looking at certain structures, such as the peritoneal brevis tendon and calcaneofibular ligament. It is helpful to perform plantar flexation of the ankle if the patient is compliant. This is an extra procedure and probably more advantageous for institutions that focus around MSK expertise, such as sports medicine institutions. Luckily, we have fellowship-trained MSK radiologists and are quite interested in optimizing our scans. There are many things that one can do to understand magic angle, but it is definitely something that MSK doctors have to work with every day as they look at tendons, ligaments, and other structures in those parts of the body."

Summary

> BY SPACE:
> - **spatial patterning:** ghosting, zippers, truncation, aliasing, moiré (via spikes), distortion (geometric), chemical shift, annefact (cusp) artifact
> - **intensity variation**: susceptibility, fat saturation inhomogeneity
> - **blurry indistinct:** motion, chemical shift (blurred edges)

> BY DIRECTION:
> - **Ghosting**: in phase encoding, a replication can be produced by pulsatile or alternating motion, such as vibrations.
>   *Remedy/improvements:* patient compliance, saturation bands, reduced likelihood of mechanical or patient vibratory motion when/if possible.
> - **Zippers**: spurious bands can be in both directions: (1) phase (most common), one manifestation is caused by not closing the door properly, or equipment emitting frequencies in room (like from pulse ox equipment); (2) frequency can happen through stimulated echoes (less common).
>   *Remedy/improvement*: close door, remove/repair offending noise-making machine/equipment.
> - **Truncation/Gibbs ranging:** can be in both directions. One anatomical area to be concerned about this is in the spine as syrinx-mimicking. Potentially other anatomy, especially on edges.
>   *Remedy/improvement:* increase matrix size in offending direction.
> - **Aliasing**: wrap around, example like seen in patella (kneecap) imaging when just one knee is to be visualized, but the knee out of view unintentionally wraps on top of the other knee in the image. Aliasing happens in phase direction.
>   *Remedy/improvement:* could be to use something like oversampling or No Phase Wrap.
> - **Wrap. herringbone (via spikes)**: waves imposed on images (dark/bright stripes), can be either direction or diagonal. Think of introducing a dot of high value in k-space and then, based on where it is, it will produce a spectral wave across the image.
>   *Remedy/improvement:* Call service to help check connectors and possible spike-detection malfunction in hardware.
> - **Distortion (geometric)**: warping due to often local variations in field (think about index of refraction with light), here local gradient fields are distorted due to bending interfaces. Affects both directions. A classic example occurs in Echo Planar Imaging (EPI) on Diffusion Weighted Imaging (DWI) and Perfusion and putting EPI phase direction in the Anterior Posterior (AP) direction which provides a more acceptable appearance (though it does not fix the distortion). Gradient-based sequences are more 'susceptible' than are spin echo-based sequences (compare the Sella anatomy on 3D-Spoiled Gradient Recalled Sequence versus Spin Echo).
>   *Remedy/improvement:* switch to another sequence, such as Spin Echo or maybe propeller—this is case-dependent. We often also use long Echo Train Length (ETL) Fast Spin Echo (FSE) sequences for helping with metal susceptibility artifacts in musculoskeletal (MSK) imaging.
> - **Chemical shift (type 1):** black interfaces because of chemical content (typically fat) with direction in frequency (primarily type 1 describes effects on spin echo type of sequences) and we note the potential spatial shift between fat and water in tissues.

Remedy/improvement: The amount of shift is due to the frequency bandwidth used. Increase the receiver bandwidth per pixel to reduce this artifact, though sometimes people read through with understanding of its presence.

- **Chemical shift (type 2):** Typically, the black interface around the organ is the one that is noted on the dual-echo sequences. At 1.5 T, ~2.25 ms TE is typically the out-of-phase and ~4.5 ms is in-phase (note this TE time is different at 3 T, which merits a separate discussion). The first of these echoes in the 1.5 T system is considered the out-of-phase. Typically, the out-of-phase provides a black line around organs, while the in-phase gets you additive signals between water and fat. Disease may occur near the boundaries between tissue types, and it is good to have both looks (in and out of phase) so you can both delineate the edge and assess the boundary of the organs.

  Remedy/improvement: In-phase and out-of-phase are used as two different 'looks' in MRI to help with evaluating disease. Be aware that one is subtractive and the other is additive. If you end up trying to look at 3 T images, then there is some more complex discussion about which echo to take first (in- versus out-of-phase) that may be useful to have with your attendings/technologist on this 3 T issue. However, in general you can follow a similar line of thinking at higher fields when using dual-echo sequences. The chemical shift artifact of the second kind only occurs with gradient echo sequences. With spin echo sequences, the 180° pulse refocuses spins to create the echo.

- **Annefact (cusp) artifact**—signal from outside of FOV (field of view) corrupts the image (either RF signal or by gradient). Shows up as blips, lines, or signal loss. The most prevalent is through the RF when you have receiver channels not properly turned off in the spine. The gradient version is because the gradients are no longer linear. For long body parts, signals from nonlinear gradients can be thrown back artifactually into the image.

  Remedy/improvement: Turn off all unnecessary RF coils when imaging (work with technologists), and/or have service check gradients performance.

- **Dielectric effects**—flip angle inhomogeneity related to RF penetration as it is absorbed differently across the patient. This leads to shading in the image. The effect becomes more prominent at 3 T versus lower field strengths. Additionally, ascites (fluid buildup in abdomen) provides a challenge to generating good RF penetration throughout.

  Remedy/improvement: Drain ascites prior to imaging when possible. Consider using less than 3 T systems if dielectric effects are suspected to be a problem.

- **Cross-talk—signal loss (and potential artifacts)** due to sequential overlapping slices. Slice selection is not perfect, and bleeding of signal between adjacent slice acquisitions can cause signal loss.

  Remedy/improvement: Check if slice interleaving has been turned on. Slice interleaving skips every other slice, then returns back to acquire the intermediate slices, allowing time enough for T1-decay to minimize signal interference between adjacent slices.

- **Inhomogeneous signal chemical saturation/particularly fat-sat**—Off-resonance refers to effects that produce shifts around the precessional frequency. Common problems have to do with susceptibility or miss-timed/aligned RF pulse excitation. The effect is to cause signal variation due to an inhomogeneous signal across the image. This is most common in the neck region of anatomy because the change in body makes it hard to get a uniform B_0 across it (thus creating off resonance effects).

  Remedy/improvement: Switch to STIR if consistent or be prepared to have to read through inhomogeneous fat saturation (places where the saturation is complete, and some places where it is not complete). Switching to alternative types of fat sat (sometimes helped in certain body parts IDEAL), improved versus classical and/or manual fat sat sometimes can 'reduce' the inhomogeneous suppression.

References

[1] Balogh E P, Miller B T and Ball J R 2015 The diagnostic process *Improving Diagnosis in Health Care* (Washington, DC: National Academies Press)

[2] Radiological image artifact (Radiology Reference Article, https://radiopaedia.org/?lang = us) (n.d.) (Retrieved 26 January 2022) https://radiopaedia.org/articles/radiological-image-artifact?lang=us

[3] Brown R W, Cheng Y-C N, Haacke E M, Thompson M R and Venkatesan R 2014 *Magnetic Resonance Imaging: Physical Principles and Sequence Design* 2nd edn (New York: Wiley)

[4] Bernstein M A, King M F and Zhou X J 2004 *Handbook of MRI Pulse Sequences* (Oxford: Elsevier)

[5] Kneeland J B, Shimakawa A and Wehrli F W 1986 Effect of intersection spacing on MR image contrast and study time *Radiology* **158** 819–22

[6] Collins C M, Liu W and Schreiber *et al* 2005 Central brightening due to constructive interference with, without, and despite dielectric resonance *J. Magn. Reson. Imaging* **21** 192–6

[7] Gabriel C, Gabriel S and Corhout E 1996 The dielectric properties of biological tissues: I. Literature survey *Phys. Med. Biol.* **41** 2231–249

[8] Phase oversampling? (n.d.). Questions and Answers in MRI (Retrieved January 26, 2022) http://mriquestions.com/phase-oversampling.html

[9] Milhorat T H 2000 Classification of syringomyelia *Neurosurg. Focus.* **8** E1

[10] Babcock E E, Brateman L and Weinreb J C *et al* 1985 Edge artifacts in MR images: chemical shift effect *J. Comput. Assist. Tomogr.* **9** 252–57 (notably one of the first descriptions of this artifact)

[11] Electron shielding 2017 Chemistry LibreTexts https://chem.libretexts.org/Bookshelves/Analytical_Chemistry/Supplemental_Modules_(Analytical_Chemistry)/Analytical_Sciences_Digital_Library/Active_Learning/In_Class_Activities/Nuclear_Magnetic_Resonance_Spectroscopy/03_Text/03_Electron_Shielding

[12] Soila K P, Viamonte M and Starewicz P M 1984 Chemical shift misregistration effect in magnetic resonance imaging *Radiology* **153** 819–20 (first published description of the chemical shift artifact)

[13] Intracranial dermoid cyst (Radiology Reference Article https://radiopaedia.org/?lang=us) (n.d.) (Retrieved March 23, 2022) https://radiopaedia.org/articles/intracranial-dermoid-cyst-1?lang=us

[14] Hood M N, Ho V B, Smirniotopoulos J G and Szumowski J 1999 Chemical shift: the artifact and clinical tool revisited *Radiographics* **19** 357–71 (excellent review)

[15] Wehrli F W, Perkins T G, Shimakawa A and Roberts F 1987 Chemical shift-induced amplitude modulations in images obtained with gradient refocusing *Magn. Reson. Imaging* **5** 157–8

[16] Fullerton G D and Rahal A 2007 Feb Collagen structure: the molecular source of the tendon magic angle effect *J. Magn. Reson. Imaging* **25** 345–61

[17] Zipper artifact (Radiology Reference Article, Radiopaedia.org) (n.d.) https://radiopaedia.org/articles/zipper-artifact?lang=us

[18] Phase-encoded motion artifact (Radiology Reference Article, Radiopaedia.org) (n.d.) https://radiopaedia.org/articles/phase-encoded-motion-artifact-3

Chapter 8

Concluding a journey through MRI

This book was a journey of connecting the dots for MRI. We began in chapter 1 with five core concepts behind MRI: waves, water and images, multiple-looks, dipoles and precession, and $\Delta E \to M$ and $\Delta M \to E$. There were also bits of physics and clinical applications intermixed to get you started. In chapter 1, you saw how MRI has unique looks that provide significant differentials for disease that other modalities may have less of a chance to achieve. We also described some limitations.

Then, in chapter 2, we discussed hardware, which is integral for technologists' understanding. This chapter included issues important for patient safety, and framed the machine as the 'camera' through which the images are derived. We introduced five hardware components: main field, RF transmitter, gradients, shim, and the RF receiver. Hopefully, you were able to gather a sense of precession of the dipole, and gain an appreciation for the brilliant discovery of the implementation of the 'gradient' by Paul Lauterbur and its advancement by Peter Mansfield, which led to the MRI revolution.

The content in chapter 3 addresses one of the most central concepts, but can also be the most challenging. Mastery of **relaxation** is important as it can help you understand the pathology of disease, as we also described the chemical moieties that your radiology attendings master. These include mineral-rich, free water, bound water, lipids, proteinaceous tissue, blood, and more, for all of which our attendings understand the appearance in MRI in great detail.

Chapter 4 mainly involved the understanding of the pulse sequence diagram, where we revealed it to you and unwrapped it in a story that hopefully aided your recall.

1) RF transmit (R = retaining wall).
2) Spear (S = slice).
3) Asp (A = acquisition).
4) Fan (F = frequency).
5) Pillar (P = phase encode).

We showed you the power and elegance of MRI, revealing to you the **spin echo**. The spin echo is a fundamental pulse sequence that is particularly useful in reducing field inhomogeneity effects that are part of the T2* mechanism. Many of today's applications in day-to-day clinical use are based on some sort of variation of incorporation of spin echo in their construction.

In chapter 5, we equipped you with three building blocks that concerned greater intuition into slice selection with the gradient and the transmit bandwidth. This was followed by more intuition on k-space, which will hopefully be helpful for your understanding of artifacts later, as well as fast sequences (not covered in this book). Finally, the elusive concept of receiver bandwidth was discussed to provide you with more tools in understanding pulse sequences

We then provided you with a model in chapter 6 to help you understand the signal-to-noise, time of sequence, and energy deposition (SAR) tradeoffs in MRI. These are essential components to your understanding that may in turn help you optimize your pulse sequence protocols and help keep your patients comfortable and safe.

In chapter 7, we covered many of the artifacts that are part of reading and creating quality images with MRI. The impact of artifacts can affect a practice. Low quality images decrease your confidence from your referring community ('referrings' such as neurosurgeons, oncologists, orthopedists, and more), as they respond to the quality of the reading as well as the images. Not only is it important for the appearance and confidence in your reads, but it can lead to false positive and false negative readings, which can be misleading and lead to downstream consequences in the treatment pathway.

Finally, the author would like to acknowledge all those who contribute to and work in the field of MRI. Over the last 30 years that he has been in the field, he appreciates all of the major contributions of applied scientists, the researchers that provide the impetus for new discoveries, and the technologists, radiologists, and referring subspecialties with whom he has worked and striven to better understand the needs for their patients. As we described in the preface, there is still more to learn, but we believe we have provided you with just enough tools for a wide variety of people to get a start with these concepts. MRI is a very significant tool in that the scanner helps bring about better healthcare. Accordingly, it is important that we continuously strive to keep up with this rapidly advancing field and work together as an interprofessional group toward enhancing our clinical teams and to achieve the highest quality of patient care.

Milton Keynes UK
Ingram Content Group UK Ltd.
UKHW051609060823
426386UK00003B/37